# LIVING DIGITAL

## 2040 | Future of Work, Education, and Healthcare

# LIVING DIGITAL

## 2040 | Future of Work, Education, and Healthcare

**Poon King Wang**
**Hyowon Lee**
**Lim Wee Kiat**
**Mohan Rajesh Elara**
**Youngjin Chae**
**Gayathri Balasubramanian**
**Aaron Yong**
**Raymond Yeong**

A project of

*Lee Kuan Yew Centre for Innovative Cities*
*Singapore University of Technology and Design, Singapore*

**World Scientific**

NEW JERSEY · LONDON · SINGAPORE · BEIJING · SHANGHAI · HONG KONG · TAIPEI · CHENNAI · TOKYO

*Published by*

World Scientific Publishing Co. Pte. Ltd.

5 Toh Tuck Link, Singapore 596224

*USA office:* 27 Warren Street, Suite 401-402, Hackensack, NJ 07601

*UK office:* 57 Shelton Street, Covent Garden, London WC2H 9HE

**National Library Board, Singapore Cataloguing in Publication Data**
Name(s): Poon, King Wang. | Lee, Hyowon, author. | Lim, Wee Kiat, author.
Title: Living digital 2040 : future of work, education and healthcare / Poon King Wang,
    Hyowon Lee, Lim Wee Kiat [and 5 others].
Other title(s): Future of work, education and healthcare
Description: Singapore : World Scientific Publishing Co. Pte Ltd, [2017]
Identifier(s): OCN 1003157137 | ISBN 978-981-11-3232-97-6 (paperback) |
    978-957-3230-70-5 (hardback)
Subject(s): LCSH: Technological innovations--Social aspects. | Work--Effect of technological
    innovations on. | Education--Effect of technological innovations on. |
    Medical care--Effect of technological innovations on.
Classification: DDC 303.483--dc23

**British Library Cataloguing-in-Publication Data**
A catalogue record for this book is available from the British Library.

Desk Editor: Philly Lim

# Acknowledgments

This research project and book would not have been possible without the support of many people. We are grateful to all of them:

- Singapore Ministry of National Development (MND) and the National Research Foundation Singapore (NRF), for funding our research through the Land and Liveability National Innovation Challenge (L2NIC).
- Lee Kuan Yew Centre for Innovative Cities (LKYCIC) and Singapore University of Technology and Design (SUTD), for supporting the research and administration.
- Economics Society of Singapore (ESS), for giving us an opportunity to co-develop an artefact of the future through their *Economics & Society* publication.
- KPMG, who opened a window to us to better appreciate the multi-faceted opportunities and challenges of digital disruption in the field of professional services and work.
- Royal Institution of Chartered Surveyors (RICS), with whom we had several interesting discussions about the future, and also co-chaired a conference "Cities in a Digital World" at the World Cities Summit.
- Singapore Technologies Endowment Program, who partnered with LKYCIC/SUTD to organize workshops about the future for the international student delegates participating in the Ecosperity Young Leaders Dialogue (Ecosperity is an annual conference organized by Singapore's Temasek Holdings).
- Victoria-Cedar Alliance (VCA), whose Imagineering Programme offered a glimpse into how they see the future.

We are also grateful to the following individuals who went out of their way to help us at various junctures in the research:

- Chow Jia Hui, Norakmal Hakim Bin Norhashim, Huang Jiayi, Madeline Wong, and Yoong Jia Jun, former SUTD students (and now alumni) who assisted with research, and provided insights into how they see the future.
- Professor Neo Boon Siong and Dr Lim Wee Kiat's colleagues at the Asian Business Case Centre, Nanyang Business School, Nanyang Technological University.
- Akshay Rao and Tan Ning, who worked with us on early research into robotics.
- Eileen Tay, volunteer editor of the Economic Society of Singapore (ESS) publication *Economics & Society*.
- Tamsin Greulich-Smith, Chief, Smart Health Leadership Centre, National University of Singapore-Institute for System Science (NUS-ISS).
- Jessica Bland and Lydia Nicholas, Nesta (formerly UK National Endowment for Science Technology and the Arts), for sharing on their experience with the use of personas and on people-powered health.
- Professor Seeram Ramakrishna, National University of Singapore, for his enthusiasm for the idea of the empathy suit and SuperCare which arose out of a workshop and subsequent discussions.
- Assoc Professor Ted Tschang, Singapore Management University, for sharing his experience and thoughts on design, design thinking, and relevant research in both.
- Kathleen Schwind, MIT sophomore who interned with the project during her exchange in SUTD.

- Ramu Uma and Aaron Poh, who interned for the project at various phases of the project.
- Malou Ko, who inspired with her dedication to design.
- Natalia Tan, who impressed with her dedication to detail.
- Assistant Professor Lyle Fearnley, Dr Corinne Ong, and Quek Ri An, from LKYCIC's Sustainable Futures project (also funded by NRF and MND through the L2NIC) and Dr Andy Zheng from LKYCIC's Smart Cities Lab (supported by the Chen Tianqiao Programme in Urban Innovation); together, we explored the impact of digital disruption and of the sharing economy on recycling tasks in the circular economy.

Across institutions and individuals, the generous sharing of expertise and time strengthened our project, and made *Living Digital 2040* better.

# Funding

This material is based on research/work supported by the Singapore Ministry of National Development and the National Research Foundation Singapore through Future of Cities project at the Lee Kuan Yew Centre for Innovative Cities, SUTD (Grant Number L2NIC-TDF1-2014-1).

Over and above this report *Living Digital 2040*, the funding also supported the following additional research/publications which informed our findings and conclusions for *Living Digital 2040*:

1) Presented
- Flexible Solar Textiles with Style for Future Smart Clothing System (Fiber Society conference 2016)

• How Professionals Are Refashioning Their Craft Around Computerization: The Case of Accountis (extended abstract accepted for poster presentation at De Lange X Conference)

• Towards Establishing Design Principles for Balancing Usability and Maintaining Cognitive Abilities (International Conference on Human-Computer Interaction, 2017)

• EduBang: Envisioning the Next Generation Video Sharing (Video Showcase ACM SIGCHI International Conference on Human Factors in Computing Systems, 2017)

2) Planning/Preparation/Work-in-Progress

• Determining Technology Trends and Forecasts of Surgical Robots by Historical Review, Bibliometrics and Comparison Analysis from 1995 to 2015

• Analysing the Innovation Growth of Robotic Pets Through Patent Data Mining

• Design Model of Wearable Smart Clothing with Multi Layering System

• How Professionals Make Sense of the Future of their Profession and Career in the Context of Increasing Automation and Digitization

• Exploring the Future Usage Scenarios and Applications Using Wearable Technologies for Healthcare

Two of our artefacts from the future were also highlighted in the public domain:

• Lesson Design Map was published in the Economic Society of Singapore's publication *Economics & Society Vol 1, 2016 — Man and Environment*. It shows how academic concepts and topics can be deconstructed and reconstructed to help students see connections between fast fashion, environment, economics and other school subjects.

- Empathy Suit was highlighted in *Tabla!* (a news publication) as an example of how smart fabrics might be used in future to create people-centered "wearables" that could help to bring people closer together.

The opinions, findings, conclusions, and recommendations expressed in this report (and additional research/publications) are those of the authors and do not necessarily reflect the views of the Singapore Ministry of National Development and the National Research Foundation Singapore.

# About the Authors

**Mr POON King Wang**, Director, Lee Kuan Yew Centre for Innovative Cities, Singapore University of Technology and Design (SUTD). He is also concurrently SUTD's Director for Strategic Planning, and Co-Director of the SUTD-JTC Industrial Infrastructure Innovation Centre. He is the Principal Investigator for the project.

**Dr Hyowon LEE**, Assistant Professor, Information Systems Technology and Design, Singapore University of Technology and Design. His field of research and expertise are in Human-Computer Interaction (HCI), interaction design, usability and user experience (UX). He is a Co-Principal Investigator for the project.

**Dr LIM Wee Kiat** was at the Lee Kuan Yew Centre for Innovative Cities as Research Fellow, and Co-Principal Investigator for the project. He is currently Research Fellow, Asian Business Case Centre, Nanyang Business School, at the Nanyang Technological University, Singapore. Prior to his PhD education in sociology, Wee Kiat worked six years in Singapore, holding research and planning positions in telecommunications and national defence sectors. His research interests lie at the intersections of organisational change, technology, and risks, focusing on crises and disasters.

**Dr MOHAN Rajesh Elara**, Assistant Professor, Engineering Product Development, Singapore University of Technology and Design. His research interests are in reconfigurable robotics with emphasis on design as well as research problems related to control, human robot interaction, navigation, and perception. He is a Co-Principal Investigator for the project.

**Dr Youngjin (Marie) CHAE**, Research Fellow, Lee Kuan Yew Centre for Innovative Cities. Marie is a research scientist in the field of technologically-intensive design for functional garments and wearable electronics. Her research interests are in the areas of creative design in smart clothing and E-textiles, energy harvesting for flexible mobile devices, emotional and humanitarian aspects of wearable devices and comfort issues of functional garments. She is a Research Fellow for the project.

**Ms Gayathri BALASUBRAMANIAN**, Research Assistant, Lee Kuan Yew Centre for Innovative Cities. Gayathri is interested in studying how the nature of human-technology interaction is evolving with rapid advancement of technology, its consequences on users' abilities, and how to design technology that inherently overcomes negative consequences of its prolonged use. She is a Research Assistant for the project.

**Mr Aaron YONG Wai Keet**, Senior Industrial Designer, Lee Kuan Yew Centre for Innovative Cities. Aaron is a practitioner of human-centered design. He conceptualizes and designs future artefacts through prototyping of future personas and scenarios. He is a Senior Industrial Designer for the project.

**Mr Raymond YEONG Wei Wen**, Research Officer, Lee Kuan Yew Centre for Innovative Cities. Raymond is interested in understanding how technology and robotics are evolving especially in the field of reconfigurable robotics, with an emphasis on design, as well as how state-of-the-art in the evolution of technology and robots affect us in our activities. He is a Research Officer for the project.

# Contents

# List of Personas and Artefacts

| | 2020 | 2030 | 2040 |
|---|---|---|---|
| **JAZ'S GRANDPARENTS** | in their late 70s | in their late 80s | in their late 90s |
| | Retired Mom & Pop Waste Collectors*<br><br>Artefact:<br>Redesigning Recycling (pg.80) | Digital Peer Tutored<br><br>Artefact:<br>SuperCare/<br>Empathy Suit (pg.167) | Digital Peer Tutors<br><br>Artefact:<br>Jazper (pg.178) |
| **JUAN** | in his late 40s | in his late 50s | in their late 60s |
| | Parent Volunteer<br><br>Artefact:<br>Lesson Design Map (pg.122) | Recycling Re-Designer<br><br>Artefact:<br>Redesigning Recycling (pg.80) | Travelling the world and mentoring students globally (when they are not taking care of Jaz's child) |
| **KIM** | | in her late 40s & 50s | |
| | | Tast Transition Expert<br><br>Artefact:<br>Task Transition Framework (pg.71) | |
| **JAZ** | in her late teens | in her 20s | in her 30s |
| | Student<br>"Volunteered" to help her father<br><br>Artefact:<br>Lesson Design Map (pg.122) | Designer<br><br>Artefact:<br>Edu-Bang (pg.129), HCI 2.0 (pg.77) | Venture Capitalist Designer<br><br>Artefact:<br>Nurture/哪吒 (Ne Zha) Bib (pg.173) |
| **JAZ'S CHILD** | | | new born |
| | | | Mommy's Little One<br><br>Artefact:<br>Nurture/哪吒 (Ne Zha) Bib (pg.173), Jazper (pg.178) |

* colloquially known as the Karung Guni

# Introduction

*"Please, no more big words."*

*"So what do we do?"*

*"We want to be hopeful."*

This book is about the impact of digital technologies on the future of work, education, and healthcare, and how we should respond. When we did the research for the book, we asked our interviewees and workshop participants what they wanted to know about this future. They told us that they wanted more than buzzwords. They wanted to know what practical steps they could take to prepare for the future. They also wanted to feel hopeful — that a dystopian digital future was not inevitable, but would instead be an opportunity to transform their communities, companies, cities, and even countries for the better.

Their heartfelt sentiments guided how we approached our research and how we published our findings. You can see this clearly in this book. It is steeped in human hopes and fears. It is underpinned by practical ideas and research rigour. We avoided big words when simple words would do. We also used visuals and stories to appeal to more readers. We believed that if more of us understood more about the economic and social impact of digital technologies, the more prepared we will be for the future.

In studying this future, we focused on work, education, and healthcare. Chapter 1 explains why. They are three human, economic, and social institutions that all of us experience every day or at seminal stages of our lives. How we tackle the impact of digital on them will substantially shape our individual and collective futures.

Chapter 2 lays out how we sought depth, definition, and diversity of how people made sense of technology and the future. It summarises our multi-method approach and how we worked with many collaborators to get at diverse and different perspectives.

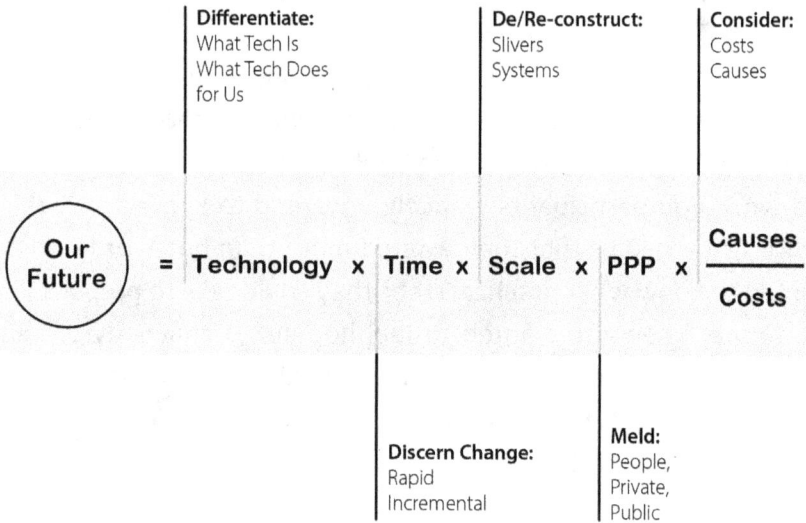

| Differentiate: | De/Re-construct: | Consider: |
|---|---|---|
| What Tech Is | Slivers | Costs |
| What Tech Does | Systems | Causes |
| for Us | | |

$$\text{Our Future} = \text{Technology} \times \text{Time} \times \text{Scale} \times \text{PPP} \times \frac{\text{Causes}}{\text{Costs}}$$

| | Discern Change: | Meld: |
|---|---|---|
| | Rapid | People, |
| | Incremental | Private, |
| | | Public |

Fig I1 — Drivers of Change

Chapter 3 distills these perspectives into five drivers of change. They help us understand how technology transforms tradition, and how tradition transforms technology. They set the stage for our subsequent chapters on scenarios, recommendations, and the future we could create for work, education, and healthcare.

Chapters 4–6 detail these futures. For each of the three areas, we show how the five drivers of change point towards scenarios that revolve around four dimensions along two axes.

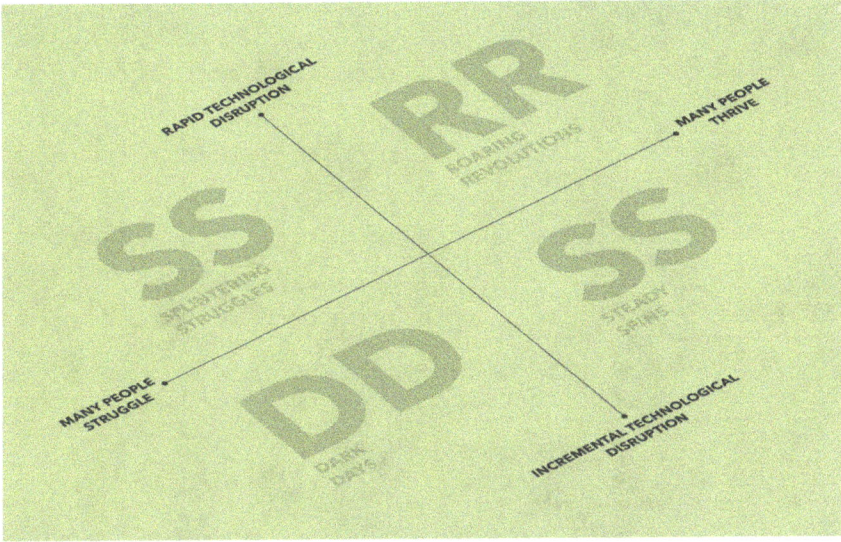

Fig I2 — Future of Work Scenarios

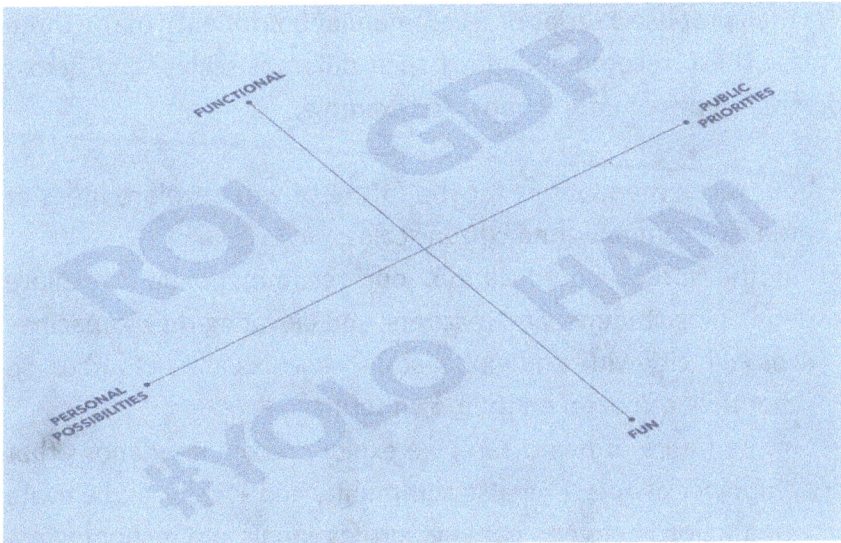

Fig I3 — Future of Education Scenarios

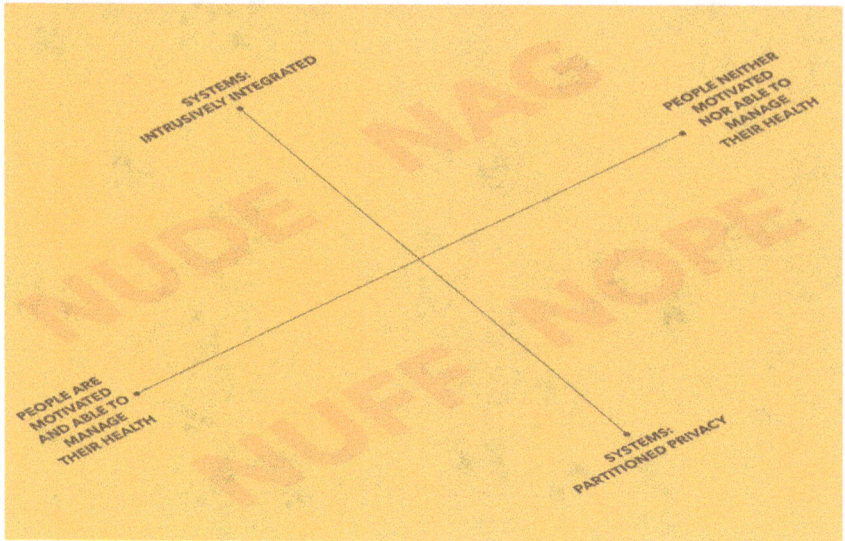

Fig I4 — Future of Healthcare Scenarios

We then propose a series of recommendations for each of the three areas. These recommendations span different scales, and across different types of interactions. For example,

- our recommendations for the future of work explore ideas at the city, company, and citizen scales;
- in the future of education, our recommendations explore equipping students and teachers, and elevating their capacities through city-wide and global-scale networks and resources, so that they can excel on their own terms;
- in the future of healthcare, we explore recommendations that empower people, energise community and society, and elevate interactions between patients, professionals and technologies.

We round out each chapter with brief narratives of what the future of work, education, and healthcare could look like if we successfully implemented the proposed recommendations.

The concluding Chapter 7 expresses our aspirations for citizens, companies, and cities. Digital innovation is an opportunity to transcend and scale across physical constraints, boundaries, and infrastructure. Through digital technologies, we can do more for each other, and more with each other. We can help each other flourish.

*Living Digital 2040* then, is a guide for navigating the impact of digital technologies on our economies, and societies, especially in the social institutions of work, education and healthcare. It points out practical ideas for what we can start doing today to prepare for tomorrow. It aims to take digital disruption and make it work for citizens, companies, and cities.

We hope it gives you hope for the future.

# 1 Beyond Physical: Cities Living Digital

## 1.1 Opening

Ulan Bator, Mongolia. Trichy, India. Choa Chu Kang, Singapore. Los Angeles, USA.

Across these four cities, through the eyes of a young boy, we glimpse our digitally-empowered future.

Adi is seven years old. He lives in Singapore, near the western end, and in public housing. He was the youngest member of Singapore's team at the 2016 Asian Youth Chess Championship held in Ulan Bator, Mongolia.

Adi loves chess. When he talks about chess, his eyes and voice sparkle. Listening to him chirp on and on about opening moves and strategies, you would not have guessed that he only qualified for the national junior chess squad a year ago. Nor that it was barely two years ago that he picked up chess, when a family friend from Los Angeles gave him a simple chess set.

How did he do it? He had a "secret weapon": the World Wide Web. Adi sparred online with top chess players worldwide, many of them older than him, and against advanced computer chess algorithms, until he was good enough to qualify for the squad.

Adi had another "secret weapon": online coaching. Adi loves his online coach, Prab. He had tried out other personal coaches, but these did not work out. It was serendipity (and a search engine) that led Adi to Coach Prab. Call it digital chemistry — through Coach Prab, Adi's skills have improved by leaps and bounds. Coach Prab is very proud of Adi's speedy progress.

Coach Prab is based in Trichy, India. He has been coaching chess players since 2005. He also suffers from double kidney failure. Because of infection risks, he is reluctant to leave his house except for dialysis treatment. He thus turned from face-to-face coaching to online coaching, so that he could continue his passion. He now coaches young chess players from over eight countries.

Coach Prab reviews the auto-recorded online games that Adi has played. He discusses where Adi did well and where he did not. Whenever he can, he also watches Adi's online games live, and discusses with him, in real time over Skype, the merits and demerits of each move. Coach Prab even uses online AI analysis to augment his feedback to Adi.

Adi loves that Coach Prab makes him think hard. Adi also loves the sessions where Coach Prab downloads Grandmaster games and explains them to him. He appreciates that he is learning immensely. He believes that he too can be a Grandmaster one day.

Adi's parents could not have imagined how all this has turned out. After all, they are not chess players themselves. They are proud of Adi and grateful to Coach Prab. More importantly, they are pleased that Adi, who used to be shy and bashful, has become more confident and articulate, and has made many friends amongst his peer competitors.

A boy in Singapore. A coach in Trichy. A competition in Ulan Bator. A friend in Los Angeles. Ignited by interest. Powered by people. Empowered by digital.

They show how cities and citizens could be *Living Digital 2040*.

## 1.2 Why *Living Digital 2040*

For the first time in human history, individuals have access to a wide array of personal, pervasive and powerful digital technologies. What is their impact on the future of cities? And what does this future hold for citizens?

## Why Living

We live digital now. In 1980, MIT Media Lab founder Nicholas Negroponte quipped that "Computing is not about computers any more. It is about living."[1]

We saw this coming with smart cities. Singapore Prime Minister, Lee Hsien Loong, for example, envisions the city-state as a Smart Nation where technologies enable people to live and connect with others better.[2]

What we have yet to see is how these technologies can change us. MIT Professor and Director of the MIT Initiative on Technology and Self, Sherry Turkle, summed this up when she said:

> We saw that we would live digital. What we have yet to see is how these technologies can change us.

> "An unstated question lies behind much of our current preoccupation with the future of technology. The question is not what will technology be like in the future, but rather, what will we be like."[3]

# Why Digital

According to the McKinsey Global Institute, "[t]oday... billions of individuals...are using global digital platforms to learn, find work, showcase their talent, and build personal networks."[4] What "was once largely confined" to advanced cities and companies is now accessible to citizens.

This transforms urban innovation. In the past, only cities and companies could access advanced technologies. They used them to build innovative solutions for urban problems. Citizen participation was limited.

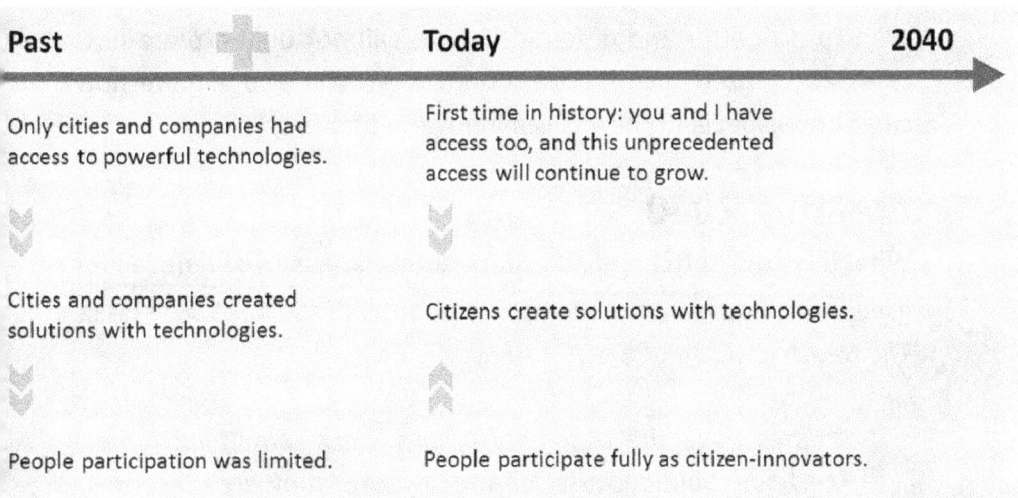

| Past | Today | 2040 |
|---|---|---|
| Only cities and companies had access to powerful technologies. | First time in history: you and I have access too, and this unprecedented access will continue to grow. | |
| Cities and companies created solutions with technologies. | Citizens create solutions with technologies. | |
| People participation was limited. | People participate fully as citizen-innovators. | |

Fig 1.2.1 — How Digital Technologies Have Transformed Urban Innovation

The advances in personal, pervasive and powerful digital technologies have changed that. Now, citizens can access advanced technologies too. They too can create innovative solutions. They can participate fully as citizen-innovators.

It is no longer just about what cities can do for citizens. It is also about what citizens can do for cities. It is about how cities and citizens can innovate together.

Smart cities are a good start. But much of the current focus is on making what already exists more efficient. It seems to be about doing better, not different.[5]

We can do better and different. So we will not only explore how cities and citizens can innovate better. We will also explore how cities and especially citizens can innovate differently.

## Creating 2040

What if we did better and different? Author and activist Jane Jacobs gave us a glimpse into the immense possibilities in her 1961 classic *The Death and Life of Great American Cities*:

> *"Cities have the capability of providing something for everybody, only because, and only when, they are created by everybody."*[6]

Five decades on, in and through digital technologies, cities have that capability, and everybody can create. Together, cities and citizens could create a very different future for cities and citizens.

It is that future we study in *Living Digital 2040*.

## 1.3 Why Future of Work, Education, and Healthcare

When we think of cities, we tend to think of physical locations, places and spaces of the city. These are tangible, visible, and often costly. They grab our attention and consume our imagination.

But the possibilities of digital suggest we need to start thinking more expansively.

Professor Michael Batty from the University College London states in his recent book *The New Science of Cities* that our understanding of cities should "no longer [be] exclusively based on theories of location".[7] As digital technologies have made possible new interactions and connections across the city and around the world,

**Digital technologies have made possible new interactions and connections across the city and around the world. Living digital means we can go beyond the physical to also focus on interactions, and explore new regional and global possibilities.**

*"we must underpin our theories with ideas about how we relate to each other... [and] switch our traditional focus from locations to interactions."*[7]

Living digital means we need to go beyond the physical to also focus on interactions. We need to go social.

We thus focus on work, education, and healthcare. Each is a social institution where many of our interactions and connections are forged. Every citizen experiences them intimately every day or at critical stages in their lives. They form a good starting point for examining "ideas about how we relate to each other",[7] and how we might go beyond physical locations.

Thinking beyond the physical and focusing on social interactions also means we can explore new regional and global possibilities. Professor Saskia Sassen of Columbia University writes in her book *Cities in a World Economy* that we can "specify a variety of transnational geographies that connect specific groups of cities" according to the social and economic processes and activities involved.[8] To her, cities offer another way to understand networks and flows of economic activity, resources, culture, and people. Because of this "proliferation of interurban networks...[t]here is no such entity as a single global city."[8]

Work, education, and healthcare also build human capabilities, as outlined in the diagram on the following page.

Human capabilities are critical: if a city had limited resources and had to choose what to invest in for the long term, building human capabilities is arguably the best bet.[10–19]

**Digital is also raising pressing questions about work, education, and healthcare. How well citizens do in them determines whether cities thrive or wither.**

But digital technologies are raising pressing questions about work, education, and healthcare.[14–19] Will jobs be created or destroyed? Will technology disrupt education and healthcare? Will technology, work, education, and healthcare narrow

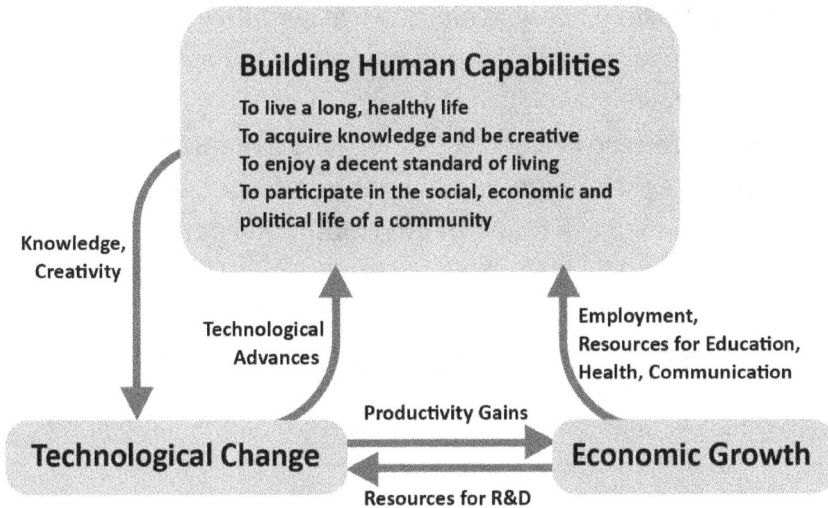

Fig 1.3.1 — Work, Education, and Healthcare Augmented with Technology - Building Human Capabilities in the City.
*Adapted from the United Nations Human Development Report 2001 - Making New Technologies Work for Human Development[9]*

or widen the gap between the haves and have-nots? How cities choose to answer these questions will determine if citizens have the capabilities to create a better future.

We focus on work, education, and healthcare for a third reason: they shape so much of who we are and what we do.

Work provides an income for us and our families. In some cases, it also endows us with the dignity of labour.

Education prepares us for work and life. It hones the intellect, seeds interests, nurtures values, builds character, and develops skills.

Healthcare makes it possible for us to live life with vigour. It ensures we have the energy and constitution to make the most of our lives.

How well citizens do in work, education, and healthcare determines whether cities thrive or wither. The future of work, education, and healthcare then, is the future of our citizens and our cities.

# 2  How and Why

## Multi-Method, Multi-Disciplinary

There are many ways to observe and analyse the world. We chose to use a multi-method approach — in-depth interviews, participation observation, group discussions, and document review — and worked with a range of collaborators (see earlier section on Acknowledgements) to understand different world views and perspectives (see Methods section at Annex A).

We also had a multi-disciplinary team with experience and expertise that includes design, sociology, human-computer interaction, human-robotics interaction, analytics, wearables, IT and organisations, fashion design, industrial design, telecommunications, banking, consumer products, and public policy.

## Choices from Voices, Voices to Choices

A report on living digital would be remiss if we did not focus on people. Harvard University cognitive scientist Professor Steven Pinker points out:

> "we have become accustomed to understanding the social world in terms of 'forces', 'pressures', 'processes', and 'developments'. It is easy to forget that those 'forces' are statistical summaries of the deeds of millions of men and women who act on their beliefs in pursuit of their desires. The habit of submerging the individual into abstractions can lead not only to bad science... but to dehumanisation."[1]

That is why we spoke to many people at length. Their voices allowed us to see things in concrete terms and not in abstractions, to reflect on the deeper meaning of their words, and to help us distil their world views, fears, and aspirations. We heard from them the choices they hoped to make, and the choices that they hoped the cities they cared about would make.

We translated these deep and rich interactions and materials into scenarios, personas, and artefacts. These give voices to the choices that cities, companies, and citizens could make.

**Speaking to many people at length allowed us to see things in concrete terms and not in abstractions. We translated their voices into scenarios, artefacts, and personas; these show us how technology transforms tradition, and how tradition transforms technology.**

Scenarios highlight the tradeoffs we have to face. Personas represent the "men and women who act on their beliefs in pursuit of their desires"[1]. Together with artefacts, they help us see and feel the future in a visceral way. As former British Museum Director Neil MacGregor puts it:

> "it's making 'things' and then coming to depend on 'things' that sets us apart from all other animals and, ultimately, turns us into the humans we are today."[2]

Things. Artefacts. Personas. Scenarios. In a study about people, living, and digital, they show us how technology transforms tradition, and how tradition transforms technology.[3–13]

## Integrated and Visual

We kept our report integrated and visual. We devoted a tremendous amount of time to uncover what really mattered. Instead of listing out all the major trends, we distilled them into just five key drivers of change, that in turn were translated into three sets of scenarios, each governed by two of the most critical considerations, centred around one simple unifying idea. Our key findings were also integrated directly into the recommendations, to show immediate links between them.

We also did not rehash topics that have already been covered extensively by other publications (such as how big Big Data is, or AI's recent breathtaking advances). In the age of digital, such information is just a mouse click or touch away (and we provide them in our References).

Finally, several sections of the report are visual in nature: we painted the scenarios, personas, and artefacts with narrative hues, rendering mental images in readers' minds. We also illustrated many of them. Making it visual helps readers grasp our ideas quickly, and they can choose to dive deeper into the text if they want to.

# 3 Drivers of Change

Technological trends and their potential impact on economy and society are well documented in many publications. Hence, we chose not to rehash what are in Six Exponential Technologies,[1] [2] Fifth Kondratiev Wave of Innovation,[3–5] Industry 4.0,[6][7] Third Industrial Revolution[8][9] and Second Machine Age.[10]

Instead, we dug deeper and explored where they converged and diverged, distilling them all into one conceptual equation, that spells out the drivers of change that will shape our future:

$$\text{Our Future} = \text{Technology} \times \text{Time} \times \text{Scale} \times \text{PPP} \times \frac{\text{Causes}}{\text{Costs}}$$

Fig 3.1 — Drivers of Change Conceptual Equation (in brief)

## Technology: Differentiate Between What It Is and What It Does for Us

We are interested in Big Data not because it is big or it is data. We are interested because Big Data can help us make decisions, optimise operations, and predict problems. We are interested in robots not because they are machines. We are interested because they can do tasks that are dull, dirty, and dangerous to humans.

In a study about the future then, it is not just what future technologies are (see Annex B for a consolidated view of selected major technology advances and forecasts), but also what they can do for us. What can digital technologies do for us? They can give us new, better, and different ways to:[11–25]

- **Live, love, learn, and earn**
Current examples: digital identities, digital health, online dating, gaming, freelancing, online courses

- **See, sense, experience, and empathise**
Current examples: augmented/virtual/mixed-reality, 3D audio, "hearables", IoT, smart fabrics, wearables, digital models, biometric and environmental sensors

- **Create, communicate, and collaborate**
Current examples: 3D printers, digital arts and design, social media, universal translation, man-machine interaction, user-generated content and crowdsourcing

- **Automate, augment, and analyse**
Current examples: robotics, autonomous vehicles, genomics (and other omics), AI, algorithms, predictive analytics, data visualisation, decision support

- **Trade and transact**
Current examples: online shopping/e-commerce, open data, sharing economy, freemiums, ad-supported, payment systems

- **Protect**
Current examples: surveillance, cyber-physical security, automated cyber defence, bitcoin/blockchain, Creative Commons, privacy, access controls, experiments with quantum communications

Understanding what technologies can do for us allows us to deeply consider what we do with them. For example, surveillance technologies can protect us; they can also be used to see what we do, intruding into our privacy. Digital identities allow us to live different "lives" for different purposes (e.g., the way teens use social media); they also make it easier to trade and transact to earn a living (e.g., World Bank's Identification for Development Initiative). It is only when we have a sense of what we do with technologies that we can begin to determine what policies and regulations are needed.

> **It is only when we have a sense of what technologies can do for us, and what we can do with them, that we can begin to determine what policies and regulations are needed, and how we can do more to improve lives.**

Moreover, as technological advances accelerate, the gap between what we currently do with technology and what we can do with it will widen (many classrooms in many cities, for example, are still far from fully exploiting the wealth of available Open Education Resources).[11] To close this gap, we need to be imaginative. We can be more imaginative and do more to improve lives by exploring what technologies can do for us, and not just what the technologies are.

## Time: Discern Between Rapid and Incremental Disruption

Disruptions happen over time, not overnight.[26–33]

In 2015, the World Economic Forum (WEF) Survey conducted a survey on game-changing digital technologies.[34] 800 executives responded to this survey. According to them, the tipping points for 21 such technologies are:

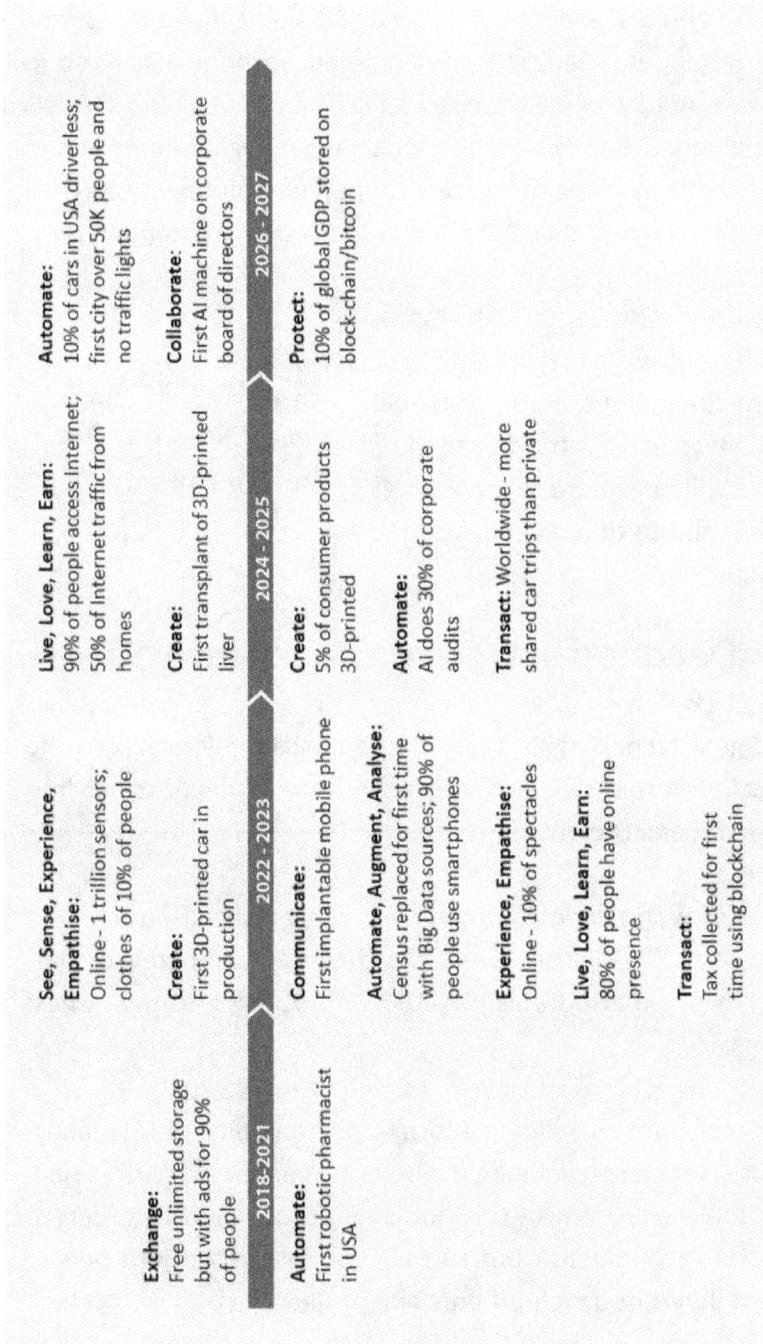

**2018-2021**

Exchange:
Free unlimited storage but with ads for 90% of people

Automate:
First robotic pharmacist in USA

**2022 - 2023**

See, Sense, Experience, Empathise:
Online - 1 trillion sensors; clothes of 10% of people

Create:
First 3D-printed car in production

Communicate:
First implantable mobile phone

Automate, Augment, Analyse:
Census replaced for first time with Big Data sources; 90% of people use smartphones

Experience, Empathise:
Online - 10% of spectacles

Live, Love, Learn, Earn:
80% of people have online presence

Transact:
Tax collected for first time using blockchain

**2024 - 2025**

Live, Love, Learn, Earn:
90% of people access internet; 50% of Internet traffic from homes

Create:
First transplant of 3D-printed liver

Create:
5% of consumer products 3D-printed

Automate:
AI does 30% of corporate audits

Transact: Worldwide - more shared car trips than private

**2026 - 2027**

Automate:
10% of cars in USA driverless; first city over 50K people and no traffic lights

Collaborate:
First AI machine on corporate board of directors

Protect:
10% of global GDP stored on block-chain/bitcoin

Fig 3.2 – Tipping Points of Game Changing Digital Technologies as Identified by World Economic Forum Survey
Adapted from World Economic Forum (WEF) Global Agenda Council on the Future of Software & Society's 2015 report
Deep Shift: Technology Tipping Points and Societal Impact 2015.[34] Illustrative - not meant to be exhaustive list of all digital technologies.

The tipping points are between 2018 and 2027: digital disruptions do not just happen. Many steps are needed. Technologies need to mature (see Annex B). Users need to be found. Business models must make sense. Behaviours need to change. Regulations need to keep up. And they all need to scale. Depending on how quickly all these happen, disruption can happen rapidly or incrementally.

Be it rapid or incremental, what matters is whether we have the foresight to discern the disruption, and the will and wisdom to act on that foresight. If we do, we will have time to respond. If we do not, all the time in the world is not enough.

**Digital disruptions do not typically happen overnight. They happen over time.**

## Scale: Deconstruct/Reconstruct into Slivers and Systems

Just as digital technologies span the nano-scale microchip to the global-scale Internet, we can use technology to deconstruct into slivers and reconstruct into systems.

Any piece of work, simple or complex, can be sliced up into the tiniest of tasks.[35–48] These slivers can then be automated or outsourced. They can also be aggregated with others into the largest of systems.

Digital models such as Building Information Modelling (BIM) make it easy to break up large-scale developments and projects into smaller tasks and pieces. These can not only be assigned to different experts and technologies, but can also be integrated into newer, larger, and different developments and projects.

In recent years, this has often been associated with the "sharing" economy. That is a misnomer. Many of the recent platforms are less about "sharing", and more about matching the supply and demand of slivers and systems.

We see the same idea of scale in education and health, and they are not limited to tasks. There are other dimensions too, such as interests, resources, causes, devices, data, and even time.

**We see the same idea of scale across work, education, and healthcare, in tasks, interests, resources, causes, devices, data, and even time.**

No interest, for example, is too small or too big. We can always find a niche interest group or resource online, or join mega global movements. Education now spans nano-degrees,[49] MicroMasters[50] and Massive Open Online Courses.

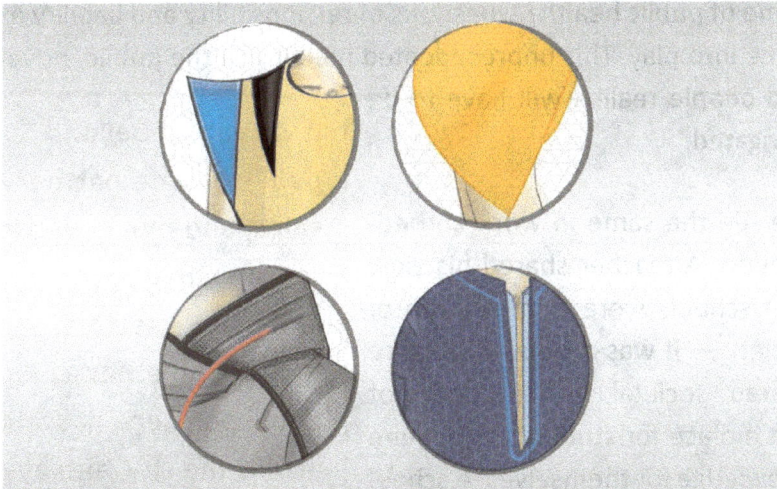

Fig 3.3 — Preventive Health
*In the future, innovations like smart fabrics equipped with sensors, data analytics, and mobile connectivity will sense and monitor our physiological signs, pre-empting and preventing illnesses (see for example Artefact from the Future: SuperCare/Empathy Suit in Chapter 6.3).*

Our genome can be separated into tiny snippets and subsequently stitched back again. We can integrate sensors, communications, IT devices, social networks, genomics, and AI into a preventive digital health system. We can measure biological signs by the micro-seconds and even decode our genes to look millennia back into our ancestry, or decades ahead into our disease risks.[51]

## PPP: Meld Public, Private, and People Sectors

Once we can deconstruct/reconstruct across scale, cities, companies, and citizens can now do much more for themselves, to each other, and for each other.[52]

In healthcare, for example, how much should be provided by the public sector, and how much of it is a personal duty? Should companies have access to our health data, or are our health data ours and ours alone? Or does it belong to the public sector in the name of public health? Questions of responsibility and liability also come into play. This unprecedented melding of the public, private, and people realms will have to be navigated.

We see the same in work and education. A teacher shared his view that schools were a microcosm of society — it was a place to nurture shared societal values, and not just a place for students to acquire knowledge for themselves. A school principal encouraged us to think about where public interests fit in as learning becomes more individu-

**If we can go beyond public-private partnerships, and find new ways for the public, private, and people sectors to work together as equal partners, at scale, across the city, and even around the world, this can be a very powerful force for the future.**

alised to personal interests, and as private companies push further into the education sector. How we choose will directly shape the experiences of our students. How they turn out will be what our workforce, citizenry, and city will look like in future.

At the same time, if we can find ways for the public, private, and people sectors to work together, this can be a very powerful force for the future. It would go beyond public-private partnerships. Look for example at how cities, companies, and citizens rallied together to help each other in recent crises. It would pioneer new models of the public, private, and people sectors collaborating as equal partners, at scale, across the city, and even around the world.

## Causes/Costs: Consider the Human Dimension

Technology has made it easier to quantify, and if something can be counted, it can be costed. The trouble is, "not everything that counts can be counted."

In the early 1900s, Taylorism, or Scientific Management, sliced up complicated tasks into easier ones for workers, and measured what workers did. For all its benefits, it was criticised as dehumanising. In the hundred years since, we have had to learn and re-learn the same lessons. The Economist recently published a piece on "Digital Taylorism".[53] It voiced concerns that as measurement technologies in the workplace become more advanced, this "modern version of 'scientific management' threatens to dehumanise the workplace."

What was missing in Taylorism was a consideration of the human cause.

What is human cause? Our interviews with professionals in health and education provide some of the best examples. We can watch the most inspiring talks online, but we are most inspired by that pat on the back from a caring teacher. We can read all we want about our health, but nothing beats that reassuring word from the doctor, nurse or dietician.

A teacher shared that the "interactions between teachers and students are the most magical, not the transference of knowledge". Patients' eyes light up when they see the doctors and nurses on their rounds. A senior nurse emphasised with great pride that when crisis hits, it is not technology but how everybody rallies together that energises each and every one of them. Another said she would identify with the profession even if she leaves it. In her own words, "once a nurse, always a nurse".

This is sometimes called the human touch. Our interviews made us realise that it might be better to think of them as human causes. Because people often commit to them, and even fight for them. They are the causes of human connection, compassion, communication, community, and conviction.

We need to consider this oft-overlooked dimension. In an interview, the head of an organisation recounted that a multi-disciplinary team picks technologies based on ROI and cost-benefit analysis. The head was taken aback when we asked if they considered what might be lost (such as these human causes), admitted they often did not, felt we were right

> Counting only what can be counted instead of what counts can be very costly. We have to remember to consider the human causes too.

to point it out and acknowledged that they should.

Counting only what can be counted instead of what counts can be very costly in our digital futures. We have to remember to consider both the cost and the cause. Especially the human ones.

## Drivers of Change in Context

It is important to remember that the five drivers of change we have just discussed are situated in context. They are part of the forces of history where work, education, and health were transformed – and continue to be transformed – by technology.

A detailed exploration of these forces is outside the ambit of this report. Instead, we have summarised them for the reader (see diagram on next page). It is largely distilled from *The World our Grandchildren Will Inherit: The Rights Revolution and Beyond* by MIT Professor Daron Acemoglu – a study "outlining the 10 most important trends that have defined our economic, social, and political lives over the last 100 years"[54] – but augmented with our project findings.[51] In fact, some of these forces go back beyond 100 years, to the Industrial Revolution.

This summary points out how different forces and trends "inter-relate".[55] They also illustrate the choices we can make that will shape our future.

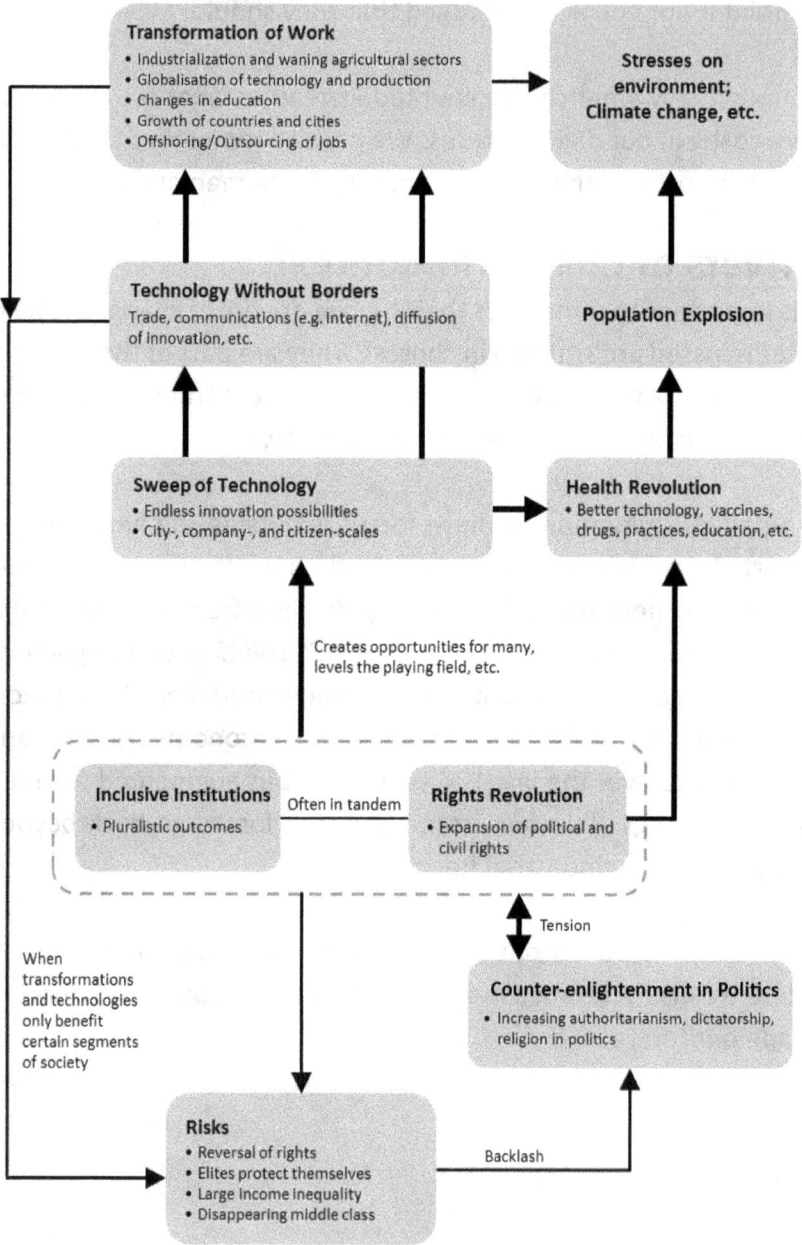

**Transformation of Work**
- Industrialization and waning agricultural sectors
- Globalisation of technology and production
- Changes in education
- Growth of countries and cities
- Offshoring/Outsourcing of jobs

**Stresses on environment; Climate change, etc.**

**Technology Without Borders**
Trade, communications (e.g. Internet), diffusion of innovation, etc.

**Population Explosion**

**Sweep of Technology**
- Endless innovation possibilities
- City-, company-, and citizen-scales

**Health Revolution**
- Better technology, vaccines, drugs, practices, education, etc.

Creates opportunities for many, levels the playing field, etc.

**Inclusive Institutions**
- Pluralistic outcomes

Often in tandem

**Rights Revolution**
- Expansion of political and civil rights

Tension

When transformations and technologies only benefit certain segments of society

**Counter-enlightenment in Politics**
- Increasing authoritarianism, dictatorship, religion in politics

**Risks**
- Reversal of rights
- Elites protect themselves
- Large income inequality
- Disappearing middle class

Backlash

Fig 3.4 — Drivers of Change Situated in Context

# Setting the Stage for Scenarios

This conceptual equation of the drivers of change distils prevailing trends, puts them in context, and sets the stage for our scenarios of the future. Tackling these drivers well amplifies what we can create for our future.

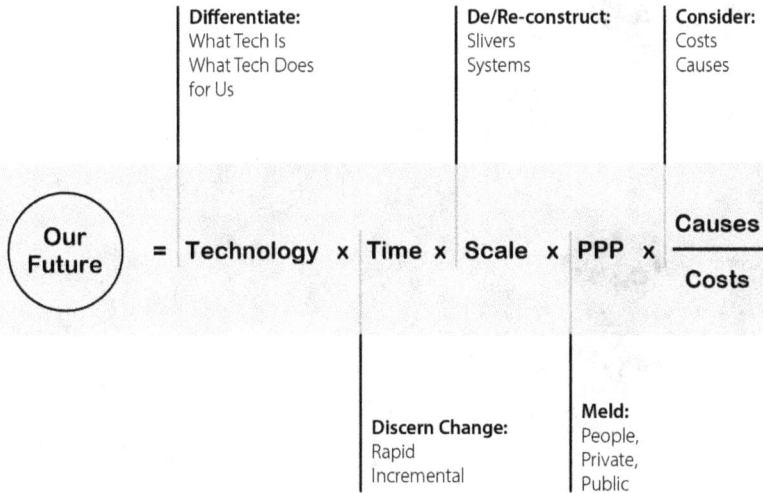

**Differentiate:**
What Tech Is
What Tech Does
for Us

**De/Re-construct:**
Slivers
Systems

**Consider:**
Costs
Causes

$$\text{Our Future} = \text{Technology} \times \text{Time} \times \text{Scale} \times \text{PPP} \times \frac{\text{Causes}}{\text{Costs}}$$

**Discern Change:**
Rapid
Incremental

**Meld:**
People,
Private,
Public

Fig 3.5 — Drivers of Change Conceptual Equation (in detail)

# 4  Future of Work

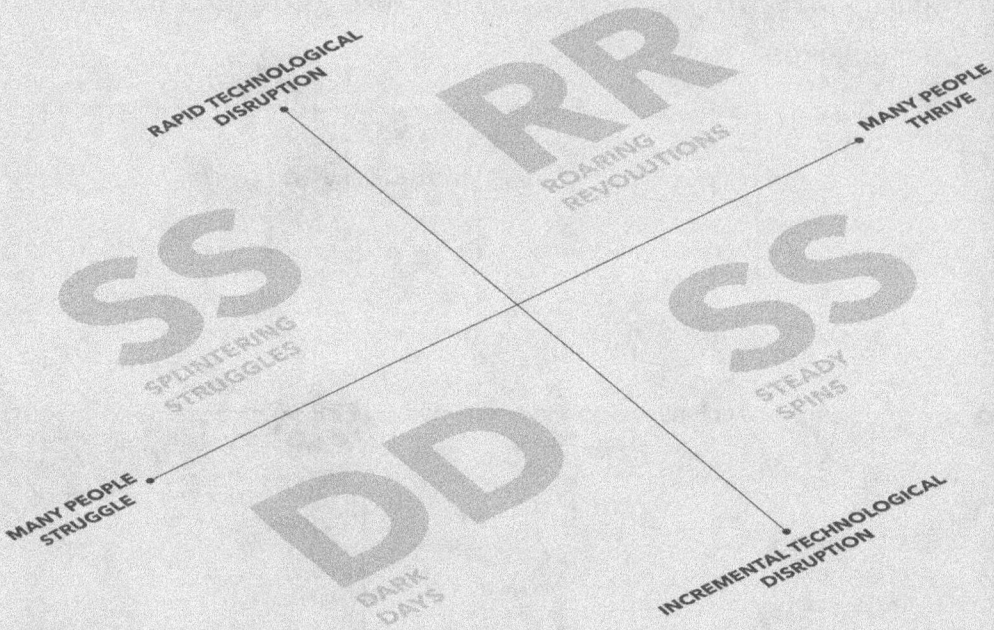

RAPID TECHNOLOGICAL DISRUPTION

MANY PEOPLE THRIVE

RR
ROARING REVOLUTIONS

SS
SPLINTERING STRUGGLES

SS
STEADY SPINS

MANY PEOPLE STRUGGLE

DD
DARK DAYS

INCREMENTAL TECHNOLOGICAL DISRUPTION

Fig 4.1.1 — Future of Work Scenarios

## 4.1 Scenarios[1–44]

Utopia. Dystopia. Optimist. Pessimist. Man. Machine.

The discourse raged on. Would accelerating advances in technology threaten jobs? And cause a great disruption in cities? Or would they save jobs? And create great abundance?

It was easy to take intellectual sides. Many did. Some even took an amusing mix of sides.

The Utoptimachs, for example, believed in a utopia of much leisure, enabled by machines of all kinds, where people would find meaning in a wee bit of work.

There was only one question that mattered to them: what do we do about the impact of technological disruption on work?

The Uto-Dystopessimachs believed in very much the same but with one crucial difference. They thought people would become hedonistic with all that leisure and pleasure, and it would all end in disaster.

And there were the Dystoptimans, who felt that a dystopian society wrecked by technology provided the most optimistic outlook. Only when man could rebuild the city and re-assert control over technology, would there be any hope of a better future.

Heady and fun stuff. But to the government leaders and company CEOs around the world, fun was the last thing on their heads. To them, it was unwise to bank on a future free from "pressing economic cares",[22] where everybody enjoyed a life of leisure. There was only one question that mattered to them: what do we do about the impact of technological disruption on work?

Answering that question determined whether their cities, companies, and citizens succeeded or failed. Lives and livelihoods were at stake. Eventually, it became clear that the answers to that question could draw inspiration from four intellectual dimensions:

1) Rapid Technological Disruption — technological innovations advancing rapidly
2) Incremental Technological Disruption — technological
innovations advancing incrementally
3) People Thrive — people are able to adapt to changes,
come what may
4) People Struggle — people struggle to adapt to changes

Where cities, companies, and citizens were along these four dimensions defined the future of work.

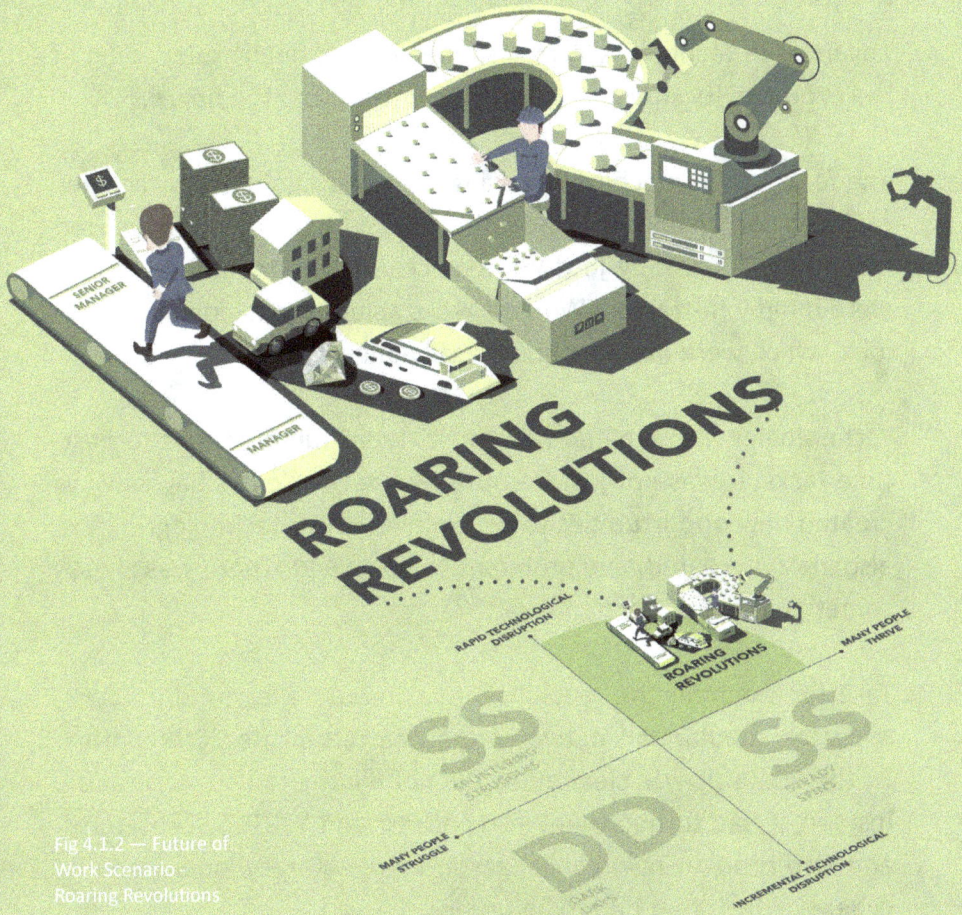

Fig 4.1.2 — Future of Work Scenario - Roaring Revolutions

## Roaring Revolutions

Citizens felt like they were on an accelerating treadmill, going round and round, faster and faster. No wonder cities and companies liked to call these rapid technological disruptions "revolutions". Life for the "Roaring Revolutions" group of cities, companies, and citizens had become one roaring rat race.

But life had also never been so good. Innovation soared. Entrepreneurship boomed. Employment exploded. The accelerating advances in technology across so many fields finally showed up in the productivity statistics, raising incomes across the board.

The fears about machine replacing man had dissipated. Machines replaced parts of jobs but not entire jobs. With the help of advanced technologies, existing jobs were quickly re-designed and workers were rapidly re-trained. Workers were able to transition into these jobs, which were often technologically-augmented.

Technologies also created new work. From full-time jobs to part-time tasks. It was not just because new technologies have always created new opportunities to be exploited: new technologies have also always created new problems to be solved. There was always something to do.

Take for example the revolution of computers, communications, software, media, and networks that started in the 20th century. By the 2010s, it was clear video did not kill the radio star after all. Instead, it had turned them into podcast and YouTube sensations, and also brought new challenges in media literacy, addiction and cyber-wellness that had to be tackled.

The "Roaring Revolutions" group of cities, companies, and citizens said it was just like history (see Fig 4.1.3). Since the Industrial Revolution, worries about the impact of technology on work have surfaced and re-surfaced several times. Each time, humans and society emerged better than worse off. There might be short-term pain but there was always long-term gain.

And the gain this time was huge. Anyone with a modicum of ambition could access global markets, funds, and resources. Success had become accessible, and very lucrative.

But this glittering prize exacted a high price. Everybody competed ferociously with one another. For the best schools. For the best jobs. And for the best careers.

It was every man for himself on a global scale. It was no longer about who could keep up, but who could keep up the fastest. If one's health suffered in the pursuit of success, there were medical technologies. If family life suffered, there were quality time family avatars. If social ties suffered, there was virtual social reality.

For the more astute cities, companies, and citizens, the relentless rat race of "Roaring Revolutions" was starting to feel like a painful marathon.

| | Industrial Revolution | Now |
|---|---|---|
| **Similarities** | Worries about unemployment caused by technological change | Worries about worker displacement caused by automation and algorithms |
| | Distress felt by those displaced | Segments of displaced workers unable to find jobs; even if they do, the jobs do not pay as well |
| | Unable to predict what new jobs would be created from the technologies then | No clear idea what new jobs might be created from the technologies today |
| | Protests and riots against changes | Occupy Wall Street; voters increasingly using ballot box to voice concerns (e.g., Brexit; populist politics) |
| **Differences** | Processes were broken down into simplified tasks that required less skill ("de-skilled" artisans) but more people | Processes broken down into different tasks that span skill levels, requiring more or less people, depending if they work full or fractional time |
| | Work and leisure often separated (start and end work at specific times) | Work and leisure blurring – by choice ("I love my work"); due to culture and communications (24x7); and shifting nature of work (e.g., freelancing) |
| **Unknowns** | More jobs were created than lost; jobs were created within existing and new sectors | Will more jobs be created than lost (many sectors already employ less people than before)? |
| | Distress: not large scale and disappeared after one generation | What is the length and scale of our current distress? |
| | Disagreements about when changes translated to higher living standards | Will changes translate to higher living standards or perpetuate prevailing inequalities? |

Fig 4.1.3 — Impact of Technology on Work (For Roaring Revolutions and Splintering Struggles)[1–23]

Fig 4.1.4 — Future of
Work Scenario -
Splintering Struggles

# Splintering Struggles

It was when society struggled and splintered that it rallied together. This was what defined the "Splintering Struggles" set of cities, companies, and citizens.

This group acknowledged that rapid technological disruption had brought tremendous improvements to daily lives. But it had caused tremendous difficulties too. This was especially so for those segments of society who struggled to adapt.

They pointed out that history was on their side too (see Fig 4.1.3). Since the Industrial Revolution, worries about the impact of technology on work have surfaced and re-surfaced several times. Even if society emerged better than worse off, there were always segments who suffered.

They beseeched society to look at the lessons of history. Society splinters whenever segments of society struggle. Left untended, the splinters become wounds that take a long time to heal.

The "Splintering Struggles" set of cities, companies, and citizens could see that these segments were growing. In the past, when disruption hit an industry, there was enough time to help affected workers adapt. They could pick up more advanced skills and move up the value chain. If they could not move up, they could at least move horizontally to another industry that had not yet been disrupted.

But the pace of technological advances had accelerated. Disruption had swept swiftly up and down within industries, and left and right across industries. It was bad enough that many workers were displaced at the same time. It was even worse that now, they could not move up or across to better or different opportunities.

The struggling workers and their companies lost faith in skills upgrading and re-training. What was the point if there was no room to move? Or if the only work available were nano-tasks with irregular payments and token employment protection? Even the fortunate ones who managed to "pivot" to another job often had to bear pay cuts. The phenomenon of training hard to pick up new skills only to pivot to a lower pay scale soon became known as the "Pyrrhic pivot".

In these trying times, cities, companies, and citizens rallied together. Part of this was altruism and care for the common man. The
struggling segments were, after all, their friends and family. But a
big part of this was pure pragmatism. I scratch your back today, so
please scratch mine tomorrow: I'm helping you now because one
day, I might struggle too, and I will need your help.

Cities stepped in to help. They responded to the calls to provide a
basic income and more social support. Companies chipped in via
co-payment arrangements. Charities and community groups re-
doubled efforts to provide financial and non-financial support. Even
citizens got into the act, volunteering to mentor those affected and
to match them to opportunities.

For now, all these could be funded by the productivity gains that
had finally materialised. Swift, widespread technology disruption
had displaced segments of society, but it had also raised productivity by leaps and bounds. These digital disruption dividends could
be shared with all.

If these gains persisted, the "Splintering Struggles" group could
continue to support those who were struggling. Would these gains
persist though? No one knew. Productivity had stalled before.

Would they be able to afford to sustain the support if that happened? Or, given the splinters that could rip through society, could
they afford not to?

# Dark Days

"Dark Days" was a deep pit no one wanted to be in. But some cities, companies, and citizens had fallen behind. They had simply failed to keep up with technological change, even though the pace of technological disruption had been incremental.

They did not plan to fall behind. So how did it happen?

The culprit was culture. And it started with success.

These cities, companies, and citizens had been successful in the past. Their way of doing things was studied worldwide. Nothing attracted adulation like success, and these cities, companies, and citizens basked in the attention.

Believing that they understood the secrets of success, they invested heavily in preserving their perceived strengths. Anything that threatened these secrets and strengths, they resisted. It was "if it ain't broke, don't fix it" on steroids.

The accelerating advances in technology hardened their resistance. Constant upgrading of technologies and skills ironically made people wary and weary of the "next big tech revolution". Privacy and cyber-physical security issues made it easy to argue about the perils of technologies instead of their potential. Democratised personal technologies put power in the hands of individuals, threatening those in power.

Over time, the prevailing culture was guilty of *argumentum ad antiquitatem*, insisting that the old way was the best way. Let's not rock the boat, they said.

All this while, the world outside was zipping past them. They failed to see that any adulation that they still received was really for past successes. Their formalised recipe for success had fossilised.

When they finally realised that they were falling behind, they started blaming each other. Cities blamed companies and citizens for not being enterprising. Companies blamed cities for not providing enough tax reliefs, grants, and incentives. Citizens blamed companies for not empowering them. And citizens and companies blamed cities for having such an "old school" education system.

It was a modern Greek tragedy, on an urban scale. Everyone wanted each other to do more, but everyone blamed each other at the same time. It would have been comical, if the consequences were not so dire.

The dark days were going to get even darker.

# Steady Spins

Life was "comfortably hard" for the cities, companies, and citizens in "Steady Spins".

It was "comfortably" because technological disruption had taken place steadily, and not suddenly as the pundits had predicted. Even though the pace of advances had increased, the pace of their diffusion had been incremental. Setting technology standards, integrating systems, reviewing regulations, establishing ethics, finding business models, and changing organisational and personal behaviours all took time. So there was time for people to adopt and adapt.

It was also "comfortably" because they had a formula for success. It was the fast follower formula: adopt what works globally, implement fast locally, and repeat religiously. The "Steady Spins" group had successfully gone through this cycle over and over again. Their fast follower formula spun like a well-oiled machine.

But it was "hard" because while fast following was easier on the intellect, it was harder on the implementation. It might be riskier and scarier to forge new paths, but no one else was on that path. But once a path had been forged, you are but one of many fast followers. Out-implementing all this competition was extremely hard work.

It was a trade-off, but one the "Steady Spins" group was prepared to accept. After all, it had paid off for so many decades. Steady economic growth, steady wage growth, and steady improvements in quality and standard of living.

Technology threatened to throw a spanner into the "Steady Spins". It was not just that technologies were creating fewer jobs over time. The economics of digital networks also meant that more and more of the rewards went to those who led, and less and less to those who followed. Over time, the "Steady Spins" cities, companies, and citizens seemed to be capturing a smaller and smaller share of the value created by new innovations.

Citizens felt it most keenly. They saw it every day in their wages and working hours. They compared themselves to their friends and colleagues in cities that were leading innovation. They saw clearly that they were working much longer hours for the same or less pay. They quipped that what "work-life balance" really meant was that work was life, or your life would be in the balance.

They wondered if there was a better way. Could "Steady Spins" keep spinning? Or was it reaching the end of its cycle?

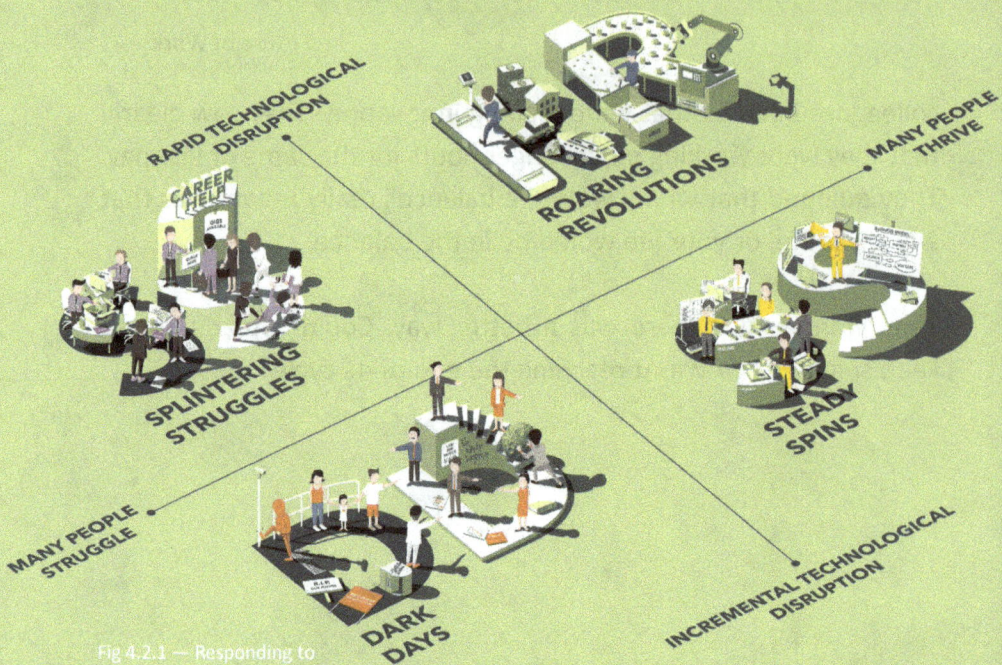

RAPID TECHNOLOGICAL DISRUPTION

ROARING REVOLUTIONS

MANY PEOPLE THRIVE

SPLINTERING STRUGGLES

STEADY SPINS

MANY PEOPLE STRUGGLE

DARK DAYS

INCREMENTAL TECHNOLOGICAL DISRUPTION

Fig 4.2.1 — Responding to Future of Work Scenarios

## 4.2 Recommendations

A city has some control over whether technological disruption occurs rapidly or incrementally, but it is only one city out of many in the world. Across the globe, cities are investing heavily in innovation, and their collective actions will determine the pace of technological disruption more than any one city's initiatives.

Helping its own citizens, however, remains the determination of a city's decisions. Our recommendations thus focus on how to help people thrive, and how to help people who struggle.

Our recommendations have **one unifying unit of analysis: tasks**.

For a long time, the unit of analysis for work was "jobs". Cities created jobs. Citizens found or changed jobs. Companies automated, outsourced or offshored jobs. But something changed in the last few decades, even as the lingo did not. Digitisation and globalisation broke down work processes and projects into tasks. These tasks were then parcelled out to computers or cities worldwide.

The atomisation of work into tasks continues unabated. It is telling that recent studies on technology and the future of work from universities and consultancies all relied on one database. That database was the USA O*NET database, which breaks occupations down into tasks (see Fig 4.2.2).

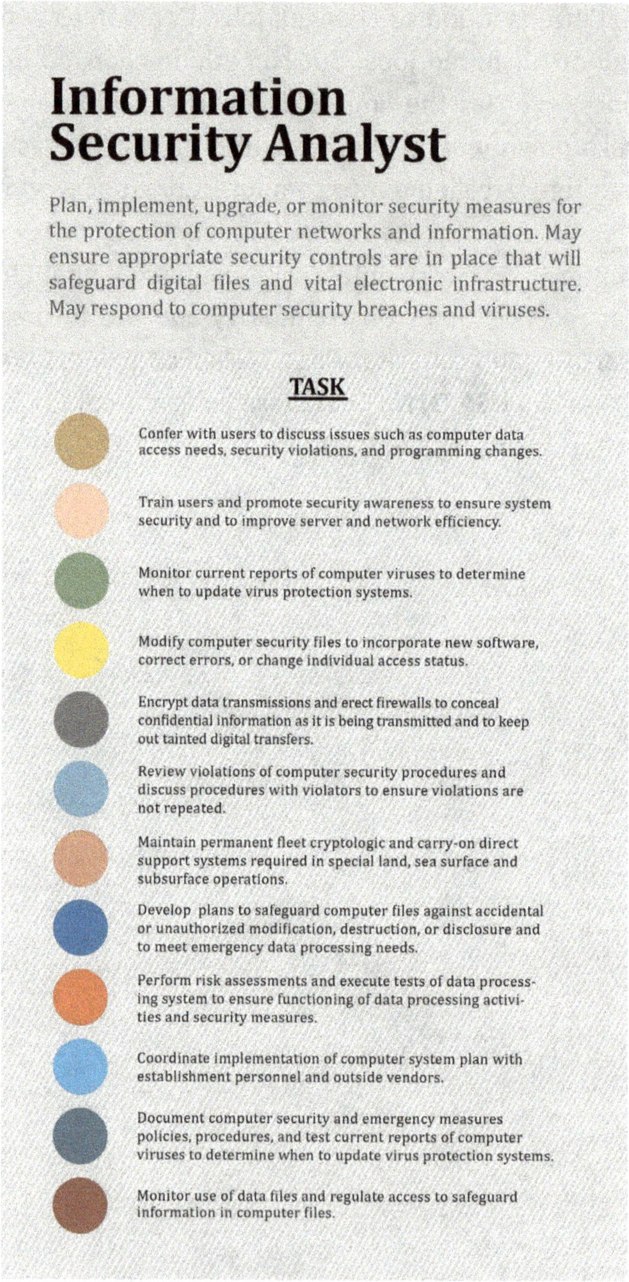

# Information Security Analyst

Plan, implement, upgrade, or monitor security measures for the protection of computer networks and information. May ensure appropriate security controls are in place that will safeguard digital files and vital electronic infrastructure. May respond to computer security breaches and viruses.

## TASK

Confer with users to discuss issues such as computer data access needs, security violations, and programming changes.

Train users and promote security awareness to ensure system security and to improve server and network efficiency.

Monitor current reports of computer viruses to determine when to update virus protection systems.

Modify computer security files to incorporate new software, correct errors, or change individual access status.

Encrypt data transmissions and erect firewalls to conceal confidential information as it is being transmitted and to keep out tainted digital transfers.

Review violations of computer security procedures and discuss procedures with violators to ensure violations are not repeated.

Maintain permanent fleet cryptologic and carry-on direct support systems required in special land, sea surface and subsurface operations.

Develop plans to safeguard computer files against accidental or unauthorized modification, destruction, or disclosure and to meet emergency data processing needs.

Perform risk assessments and execute tests of data processing system to ensure functioning of data processing activities and security measures.

Coordinate implementation of computer system plan with establishment personnel and outside vendors.

Document computer security and emergency measures policies, procedures, and test current reports of computer viruses to determine when to update virus protection systems.

Monitor use of data files and regulate access to safeguard information in computer files.

Fig 4.2.2 — The USA O*NET Breaks Down Occupations into Tasks (in this Example, 'Information Security Analyst' is Broken Down into 12 Tasks)

Hence, if we want to understand the future of work, we need to have a finer and more accurate unit of analysis than "jobs". We need to analyse "tasks".

Our recommendations thus explore tasks at different scales: city, company, and citizen.

| Scale | Recommendations |
|---|---|
| **City-scale** | 1) Explore task-based analysis of the city's economy: keep pace with what digital technology is doing to work. |
| | 2) Redesign work in existing sectors: start with what tasks different people are good at. |
| | 3) Re-imagine work and school: erase the line between them, map and match workers and students better, and build new collaboration and business models. |
| | 4) Create an improved O*NET-type database: help government agencies, companies and  citizens master tasks. |
| **Company-scale** | 5) Help employees transition: help them upgrade skills and find new work. |
| | 6) Assess enterprise risk to disruption: become more resilient. |
| | 7) Expand expertise networks and strategies: re-think and re-organise work. |
| | 8) Re-imagine new kinds of work: ensure technologies work with people, not against them. |
| **Citizen-scale** | 9) Take displacement and disruption to task: expand options for finding new work and for skills upgrading. |

Fig 4.2.3 — Summary of Recommendations for Future of Work - City-scale, Company-scale, and Citizen-scale

# City-scale

**1) Explore task-based analysis of the city's economy: keep pace with what digital technology is doing to work**

MIT Professors Daron Acemoglu and David Autor have pointed out that current models of analysis "do not include a meaningful role for 'tasks'".[45] Professor Autor subsequently wrote that "[a] growing body of literature argues that the shifting allocation of tasks between capital and labour—and between domestic and foreign labour—has played a key role in reshaping the structure of labour demand in industrialised countries in recent decades".[46]

> Research has found that tasks have "played a key role in reshaping the structure of labor demand" in many cities, and that "cities that have higher shares of connected tasks experienced higher employment growth."

Technology trends corroborate with the focus on tasks. In the debate about technology and the future of work, it is easy to overlook the fact that advanced technologies do not replace an entire job all at once.[47–49] In the field of artificial intelligence (AI) for example, what we have today is "narrow" AI, and not "general purpose" AI. Stanford University's report "Artificial Intelligence and Life in 2030" states clearly that "[i]n the approaches the Study Panel considered, none suggest there is currently a "general purpose" AI. While drawing on common research and technologies, AI systems are specialised to accomplish particular tasks, and each application requires years of focused research and a careful, unique construction."[50]

In other words, what we have is "narrow" AI. It replaces physical and cognitive tasks bit by bit.

Technologies and cab drivers are a good example. The GPS first replaced some of the drivers' cognitive tasks in navigation and routing. Apps like Uber and Grab then automated some of the tasks associated with hailing a ride and picking up a passenger. In the future, autonomous vehicles could replace the driving tasks of cab drivers.

Technology thus replaces individual tasks over time. When enough tasks within a job are replaced, then the worker with that job is displaced. The consultancy McKinsey concluded similarly in a July 2016 report that "[a]nalysing work activities rather than occupations is the most accurate way to examine the technical feasibility of automation."[51]

Hence, if we explore the economy as a collection of tasks, we could have the granularity to see which tasks (and subsequently jobs and companies) will be disrupted, and how soon. We can thus better prepare industry clusters, companies, and citizens for the future.

At the same time, task-based economic analysis can pinpoint opportunities for future employment growth. When the Netherlands Bureau for Economic Analysis conducted a study on "Cities, Tasks, and Skills", they found that "cities with higher shares of connected tasks experienced higher employment growth".[52]

A study of 168 U.S. cities showed "a positive correlation between task connectivity and subsequent employment change, which suggests that more connected tasks have gained in terms of employment share over the last two decades" (emphasis added).[53]

They also found that "spatial <u>connectivity</u> between tasks seems to be more effective than spatial <u>concentration</u> of certain tasks and labour pool suitability to explain employment growth" (emphasis added). They summarised that "task connectivity explains a significant part of the changes in employment", which alternative measures "such as the spatial concentration of tasks, do not explain", and that this "importance of [task] connectivity... is not limited to either some manufacturing or services industries or particular skill groups."[54]

Focusing on the granular level of tasks is a comparatively recent approach. The early evidence suggests several benefits. We are better able to assess the pace of digital disruption, understand shifts between capital and labour (domestic and foreign), see how activities in cities are connected, and potentially grow employment.

**2) Redesign work in existing sectors: start with what tasks different people are good at**

The division of labor between man and machine will shift over time. Tasks give us the resolution to rethink and redesign. We can see where and how technology can automate, substitute, augment, and complement man in concert.

Focusing on tasks also means we can design work according to the tasks people are good at, instead of what jobs they are good at. It is a subtle distinction, but one that can make a big difference.

Take the family doctor (also known as the GP - general practitioner). Opinions across our interviews (from medical professionals, to think tanks, to entrepreneurs) were divided about their future. Some thought that with advanced screening technologies and search engines, people would no longer need to visit GPs. Others thought that with advanced technologies, GPs could take on specialist diagnosis. They could also catch conditions that machines miss. Another group felt GPs would remain critical because people need a doctor's reassurance and empathy, especially in an emergency. GPs could also consider related cultural and religious sensitivities, or family issues the patient might be facing. And a last group reimagined GPs as health and well-being planners, responsible for keeping a community healthy.

The opinions differ because they focus on different tasks. Advanced technologies could take on many of the basic screening and monitoring tasks. But you would still likely need a doctor to help you make sense of the data and information (one interviewee said there is just too much information out there, and they often conflict), and to make a comprehensive analysis. The trust we have in a doctor's expertise lies behind why we prefer to be reassured by them, and why we think that trust could be a powerful force for community well-being.

By looking at tasks, we have a better idea of what can and cannot be replaced by technology, and why. We can also tell where technology complements or substitutes the doctor's work. We can then reconfigure the combination of tasks and technologies. Tasks give us the resolution to rethink and redesign for the better.

The division of labour between man and machine over these different tasks will of course shift over time with the advances in digital technology. But that only reinforces why it is important to understand the shifts and opportunities at the task-level.

This approach works for different sectors and profiles of workers.

For example, the elderly are a frequent sight at food centres. Many work as cleaners, clearing the tables after customers have polished off plates of food. It can be a tough physical task, and arguably unsuitable for the elderly. One idea on how to redesign the food centre tasks for the elderly came up during one of our discussions with informants. In a society where the elderly is often accorded respect, they could instead take on tasks where they could use that respect. If they are given the role of a Hygiene Supervisor, for example, this respect could be used to "nudge" customers to clear their own trays. Consequently, the technologies that need to be developed might then be those that make it convenient for customers to do this dirty and dull task themselves, instead of robotics that replace the elderly completely.

The special needs sector provides another example. Two interviewees in this sector suggested that "if companies could tell us the tasks needed in addition to the skills, it would make training more worthwhile". People with autism can be better matched according to their strengths to more opportunities,[55] and they would also better understand how the skills they are learning can be applied across different settings.

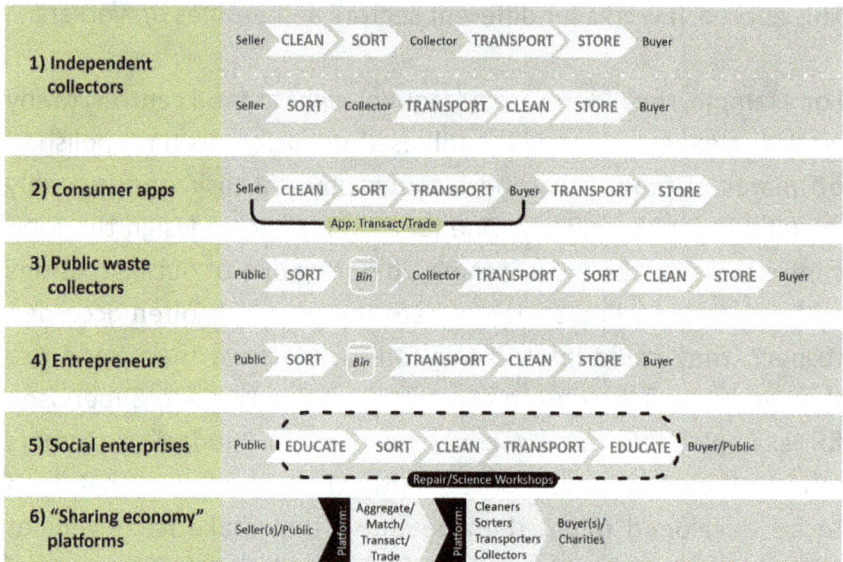

Fig 4.2.4 — Redesigning Recycling and the Circular Economy (see also Artefact from the Future in section 4.3)

A last example is Redesigning Recycling in the Circular Economy. This is an ongoing collaboration with two other projects in the Lee Kuan Yew Centre for Innovative Cities: Sustainable Futures (also funded by NRF and MND through the L2NIC) and the Smart Cities Lab (supported by the Chen Tianqiao Programme in Urban Innovation).

The collaboration breaks down the chain of activities typically found in recycling into its component tasks, such as collecting, cleaning, sorting, transporting, storing, and buying. By re-configuring these tasks, we identified ways to redesign the entire chain. These in turn could improve existing processes and create opportunities for technological innovators, entrepreneurs, and social enterprises (see Artefact from the Future — Redesigning Recycling for the Circular Economy in section 4.3).

Across ages and abilities, and from segments to sectors, focusing on what tasks people are good at creates new opportunities to be imaginative about how we can redesign the future of work.

**3) Reimagine work and school: erase the line between them, map and match workers and students better, and build new collaboration and business models**

If work is being deconstructed into tasks, we can begin to educate and train workers and students differently. For example, employers can specify the tasks that are critical to their industry. We can draw connections between these tasks and the modular concepts and topics in different courses' curricula (see Future of Education - Lesson Design Map). A lesson and training plan, with relevant content, can then be automatically composed and combined into a lesson (see Future of Education - EduBang) that is delivered in class or online to the workers and students.

Using tasks to erase the line between work and school intensifies school's relevance to work, and work's relevance to school. The partnership between cities, companies, and citizens will intensify as a result.

Do this for enough tasks, and the line between work, training and school has disappeared. When workers and students complete their projects, assignments and exams for these lessons, they will effectively be graded on how well they understand and can apply the classroom concepts to actual work settings. Employers will thus be able to match work needs better: the good grades map directly to their work requirements. And if two workers or students have the same grades but in different tasks, employers will develop a more sophisticated appreciation of the similarities and differences between them.

We can even conduct the classroom training as a training simulation. With advances in graphics and virtual/mixed/augmented reality, a wide range of immersive simulated environments can be created for a wide variety of tasks. These can be used to give practical hands-on training and assessment for workers and students. They will be able to develop strong academic, applied, and hands-on skills. Furthermore, the training can be tailored to train the right combination of academic, applied, and hands-on skills, according to the strengths and needs of each worker or student.

Erasing the line between work and school will intensify the partnership between government agencies and companies. Intensifying school's relevance to work means companies might be prepared to set aside part of their training budgets to fund programs in schools. This can help fund some of the costs needed to build new task-based systems and infrastructure (such as training simulations). Intensifying work's relevance to school means schools have systems and infrastructure that companies could use. This reduces costs for businesses. Either way, new and sustainable business models can be created to fund the required technological investments.

**4) Create an improved O*NET-type database: help government agencies, companies, and citizens master tasks**

As highlighted earlier, many recent studies used the USA O*NET database. This attests to the value of a task-based database. It also suggests we need to be mindful of the findings' transferability to other cities and countries.

If cities want to build new models to study the economy, redesign sectors, and match work and education better, they should build their own versions of tasks databases that also take advantage of recent advances in data collection and analysis.

If cities want to build new models to study the economy, redesign sectors, and match work to training and education better, it would be useful to have their own version of the O*NET. They should build on what makes it valuable, but also address its shortcomings.[56][57]

An improved O*NET-type database could have the following:
- Weighting the importance of different skills to different tasks and occupations
- Weighting the importance of different tasks and skills to different industries
- Updating – on a regular basis – emerging technologies that are complementary to the combination of tasks and skills
- Including tasks that are performed solely by technology (gives a complete picture of all tasks i.e., in addition to those done by humans only, and those augmented by technology)
- Determining how connected different tasks are within and across companies, industries and even geographies

An improved O*NET-type database should also take advantage of recent advances in Big Data, IoT and online career networks. These could help make the database more accurate and more comprehensive. It might also lower the costs of building, maintaining and updating it.

## Company-scale

### 5) Help employees transition: help them upgrade skills and find new work

Companies can focus on tasks to better help their employees transition to new tasks and work, whether those workers are displaced or are looking to upgrade. They can help employees transition through a process like the Task Transition Framework summarised in figures 4.2.5 and 4.2.6 (see section 4.3 – Artefact from the Future for full explanation).

### 6) Assess enterprise risk to disruption: become more resilient

Our interviews took place in the midst of the raging debate about how technology might or might not displace work. While some experts warned of impending dislocation, other experts echoed what Professor Richard Susskind, IT Adviser to the Lord Chief Justice of England and Wales, and Daniel Susskind wrote in their book *The Future of the Professions*:

> "... *increasingly capable machines, operating on their own or with non-specialist users, will take on many of the tasks that have been the historic preserve of the professions... an 'incremental transformation' in the way that we produce and distribute expertise in society.*"[58]

# TASK TRANSITION FRAMEWORK

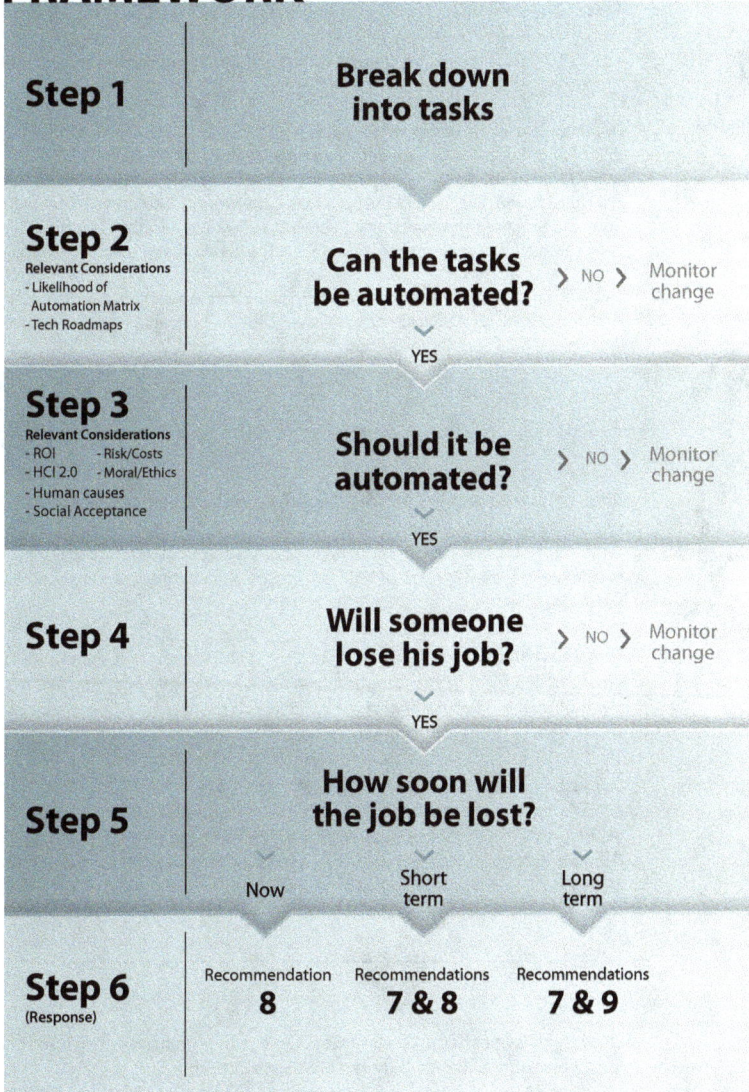

| | |
|---|---|
| **Step 1** | **Break down into tasks** |

**Step 2**

Relevant Considerations
- Likelihood of
  Automation Matrix
- Tech Roadmaps

**Can the tasks be automated?** ❯ NO ❯ Monitor change

⌄ YES

**Step 3**

Relevant Considerations
- ROI        - Risk/Costs
- HCI 2.0   - Moral/Ethics
- Human causes
- Social Acceptance

**Should it be automated?** ❯ NO ❯ Monitor change

⌄ YES

**Step 4**

**Will someone lose his job?** ❯ NO ❯ Monitor change

⌄ YES

**Step 5**

**How soon will the job be lost?**

| Now | Short term | Long term |
|---|---|---|

**Step 6**

(Response)

| Recommendation | Recommendations | Recommendations |
|---|---|---|
| **8** | **7 & 8** | **7 & 9** |

Fig 4.2.5 — Task Transition Framework (see expanded diagram in section 4.3 — Artefact from the Future)

| | |
|---|---|
| **Step 1**<br>**Break down into tasks** | Break down what workers do into tasks, and assess if these can be automated:<br>• tasks that are routine, predictable, and codifiable will likely be automated in the short term |
| **Step 2**<br>**Can the tasks be automated?** | • tasks that are non-routine, unpredictable, and need flexibility and judgement will take longer<br><br>Our interviews with workers across different trades and professions confirmed this. In auditing and accounting, for example, auditing which tends to be rules-based might see more automation than accounting which tends to be more principles-based. Even so, across both fields, the work of interpreting new regulations and practices and how those match up against the peculiar circumstances of each company's operations remain firmly in the hands of humans.<br><br>This will only change if there are game-changing innovations that overcome current technological limitations such as perception and manipulation, creative intelligence, and social intelligence. [58] [59] [60] |
| **Step 3**<br>**Should it be automated?** | Decision-makers should consider the Return of Investment (ROI), social readiness/acceptance (e.g. a hospital head told us that human acceptance remains the biggest barrier to using robots in surgeries), and the value of human causes (see Drivers of Change) before automating tasks that can be automated.<br><br>They should also assess the risks and costs of the automation, especially of errors arising from the automation (such as a false negative or positive of a medical test, or the loss of property and lives if there is an autonomous car collision).<br><br>Our last criterion is peculiar to the digital age. Decision-makers should assess if automating the process will affect long-term cognitive capabilities to build expertise. As a result of our research, we coined the term "HCI 2.0 (Human-Computer Interaction 2.0)" to account for this criterion (see Chapter 4.3 – Task Transition Framework). |
| **Steps 4 - 6**<br>**How do we help those who lose their jobs?** | Once they have considered all these, they can determine who will lose their jobs, how soon, and how they can help each of them. |

Fig 4.2.6 — Steps in Task Transition Framework (see also Artefact from the Future in section 4.3)

Our interviewees also shared similar sentiments. They echoed Pew Research Centre's *Public Predictions for the Future of Workforce Automation*: they did not expect to be substantially impacted immediately.[59]

Are the former experts too dystopian? Or are the latter professions and companies too semi-utopian?

The polarised views demonstrate that it depends on which tasks are being considered. Companies can thus assess their enterprise risk according to the tasks they and their employees do. Piece them all together and a more comprehensive, realistic and robust picture of their companies' risk to disruption will emerge. They can then take the necessary measures for their companies to weather or ride the waves of disruption.

**7) Expand expertise networks and strategies: re-think and re-organise work**

Thinking in terms of tasks also offers an interesting new opportunity: exploring innovative new ways to re-organise work.

At the heart of excelling at any task is expertise. Expertise, when broken down, is essentially an agile combination of technologies (broadly defined to encompass tools and techniques), attitudes, knowledge, and skills. For brevity, we call it TASK (Technologies, Attitudes, Skills, Knowledge). The better we are at combining TASK to solve problems or pursue opportunities, the greater our expertise.

In the past, TASK mostly resided with established experts. They were the professionals, the craftsmen, and those with the requisite qualifications, certifications and training. If we wanted something done, and we wanted to be sure, we would go to these experts. More often than not, the majority of the experts and expertise were sourced locally.

# 4D of TASK

TASK (Technologies, Abilities, Skills, Knowledge) has been Disrupted, Deconstructed, Diversified and Democratised

## local
### Established Experts

(e.g. professions, formally trained occupations, certifications etc.)

## PAST MODEL

local . regional . global
### Established Experts

**1** Experts (doing it the way it has always been done)

**2** Experts (aided/augmented with advanced technologies)

**3** Networks/communities of experts (e.g. medical specialists, flash teams)

# 9 EXISTING & EMERGING MODELS DISRUPTED BY DIGITAL

local . regional . global
### New & Non-Traditional Experts

**4** Experts (with non-traditional qualifications e.g. nano-degrees)

**5** Para-professionals (often aided/augmented with advanced technologies)

**6** Users (augmented with smart self-help systems, and/or with a network/community of other users)

local . regional . global
### Machines & Algorithms

**7** Fully automated (based on human-designed parameters but no human intervention e.g. automated alerts, IoT)

**8** Machine generated (TASK originates from machines e.g. deep learning)

**9** "Turing technologies" (i.e. performs like an autonomous human, we can't tell difference)

Fig 4.2.7 — Deconstruction, Democratisation, and Diversification of TASK (Technologies, Abilities, Skills, Knowledge)[59–71]

Digital technologies disrupted that modus operandi. It democratised, diversified, and deconstructed TASK. Our understanding of what counts as an expert and where experts can come from has expanded in the following ways:

> Both what counts as an expert and where experts can come from have expanded. From task-scale to global-scale, we can take advantage of this expanded access to technologies, abilities, skills, and knowledge.

a) Established experts: While we still have the established experts as before, the way they operate has broadened. Some will continue doing it the way they always have (either because their experience is indeed superior to the available technology, or because they are just more comfortable with how it has been done). Others would augment their capabilities with advanced technologies. And others would join forces, and work together as a network or community of experts and specialists.

b) New experts: A new group of non-traditional experts have also emerged. They might have a different set of qualifications; or they might be para-professionals, who can now do what used to be the established experts' work because they have better training or are augmented with advanced technologies. They could also be users, who, armed with self-help technology systems and the support of like-minded users, are able to offer high quality expertise.

c) Machine experts: Expertise can now also be found in fully automated systems. With deep learning and advanced artificial intelligence, machines might even generate their own expertise. In time to come, if technologies ever pass or come close to passing the Turing test, they might take over the role of the human expert completely.

Because digital technology has broken down work and jobs into slivers and slices, there are now many more tasks and many more ways to access TASK. They can now be performed by many more experts, and more types of experts. And they can be found in many more geographies. We are no longer limited to what we can find around us.

With digital technology and globalisation, we can go from task-scale to global-scale. If companies can innovate and take advantage of this expanded access in scale and scope of TASK, and also help their workers transition and upgrade (see earlier recommendations), they will be better positioned for the future of work.

**8) Reimagine new kinds of work: technologies working with people, not against them**

A lot of the prevailing discussions on technology and work focus on whether jobs will be lost or not. Not as much discussion has been about what new jobs – and work in general – could be generated. This is likely because losses are where impact is most keenly felt. At the same time, we have been notoriously bad at predicting what new jobs – and work in general – new technologies can create (see Fig 4.1.3 in Future of Work Scenarios).

Recommendations 2, 3, 5, and 7 all suggest that there are many new possibilities for creating new opportunities. Technologies have become so powerful that as much as they can substitute what humans are doing, we have to remember that they can also complement what humans are doing. MIT Professor David Autor writes that "[t]asks that cannot be substituted by automation are generally complemented by it."[72]

Partly because we have been bad at it, we do not spend enough time imagining how we can design technologies to complement the human tasks that cannot be substituted. When technologies give us the ability to either substitute or complement, it means we have a choice. And if we wish to exercise that choice wisely, we need to give it thought and to experiment with new possibilities.

Startups do this by pioneering new types of work. But this issue is too important to be left to just startups. Larger companies should also be thinking and experimenting with them.

One way to do this is via integration into their existing work. Larger companies are already experimenting with collaborating with start-ups and innovation ecosystems. These could evolve into new work and new ways of working.

Another way is to do this through existing Corporate Social Responsibility (CSR) strategies.[73] If one of the greatest social challenges of our time is the economic questions surrounding technology and work, companies should consider dedicating considerable CSR resources working with their partners, government agencies and even schools to re-imagine how to design work and technologies

We do not spend enough time imagining how we can design new work and new technologies that complement the human tasks that cannot be substituted. We can do this in startups, in large companies, collaboration eco-systems, and through Corporate Social Responsibility strategies.

such that they work with and for people, and not against people.

# Individual-scale

**9) Take displacement and disruption to task: expand options for finding new work and for skills upgrading**
A task approach is also helpful to workers who are displaced or about to be displaced by technology. They can expand their options about what they could do next by examining related tasks. They are likely to be able to find new work more easily as a result.

Take for example the job of an Information Security Analyst. This is currently a growth area but there are now major initiatives to try to automate many of the cyber-security tasks that are now performed by humans (see for example DARPA's Cyber Grand Challenge, which aims to spur innovations in autonomous protection and attack of systems).[74][75] In time to come, Information Security Analysts might be displaced too. What can they do when that happens?

Fig 4.2.8 shows how tasks can help. The information is drawn from the USA O*NET database. Besides detailing the various tasks associated with each occupation, it also identifies the links to other occupations that share similar tasks.

# Information Security Analyst

Plan, implement, upgrade, or monitor security measures for the protection of computer networks and information. May ensure appropriate security controls are in place that will safeguard digital files and vital electronic infrastructure. May respond to computer security breaches and viruses.

**Related job titles:**

- Software Quality Assurance Engineer
- Computer Systems Manager
- Clinical Data Manager
- Computer and Information Research Scientist
- Computer User Support Specialist
- Computer Network Architect
- Computer Programmer
- Document Management Specialist
- Database Administrator
- Computer Network Support Specialist
- Telecommunications Engineering Specialist
- Web Developer
- Network and Computer Systems Administrator
- Informatics Nurse Specialist
- Web Administrator

## TASK

- Confer with users to discuss issues such as computer data access needs, security violations, and programming changes.
- Train users and promote security awareness to ensure system security and to improve server and network efficiency.
- Monitor current reports of computer viruses to determine when to update virus protection systems.
- Modify computer security files to incorporate new software, correct errors, or change individual access status.
- Encrypt data transmissions and erect firewalls to conceal confidential information as it is being transmitted and to keep out tainted digital transfers.
- Review violations of computer security procedures and discuss procedures with violators to ensure violations are not repeated.
- Maintain permanent fleet cryptologic and carry-on direct support systems required in special land, sea surface and subsurface operations.
- Develop plans to safeguard computer files against accidental or unauthorized modification, destruction, or disclosure and to meet emergency data processing needs.
- Perform risk assessments and execute tests of data processing system to ensure functioning of data processing activities and security measures.
- Coordinate implementation of computer system plan with establishment personnel and outside vendors.
- Document computer security and emergency measures policies, procedures, and test current reports of computer viruses to determine when to update virus protection systems.

11 · 10 · 9 · 8 · 8 · 8 · 8 · 6 · 6 · 6 · 4

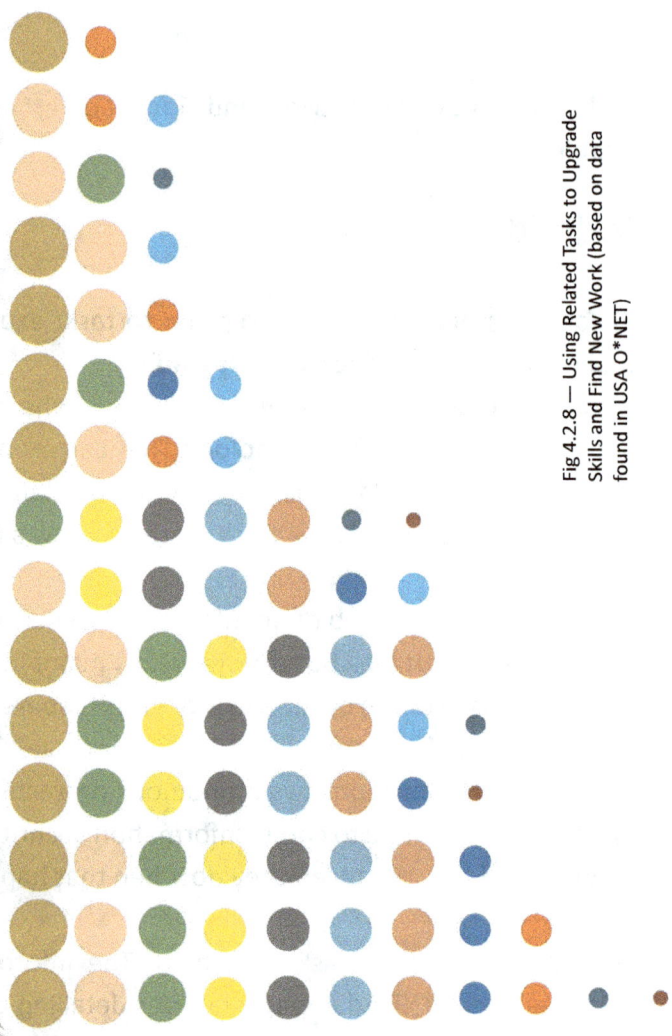

Fig 4.2.8 — Using Related Tasks to Upgrade Skills and Find New Work (based on data found in USA O*NET)

Looking from left to right, the displaced Information Security Analyst can easily identify adjacent occupations which value the tasks he has experience in. By looking at tasks, the options have expanded: there are opportunities in his current sector and in adjacent sectors.

Workers can also identify which adjacent occupations they could move to and which value the tasks they have experience in. Thus, by looking at tasks, the options have expanded: there are opportunities not only within their current sectors, but also in adjacent sectors.

Workers can also go from the right to left. This will be useful for workers who are in occupations that are not immediately threatened, but know it is a matter of time. One of the challenges of upgrading skills ahead of time is the nagging question of whether those skills are worth the investment in time and money. Taking a task approach means the worker now knows what specific tasks to target, making it easier for him to pinpoint exactly what skills and new technologies need to be picked up.

Take for example the Computer Programmer (see Fig 4.2.8). His job might be threatened by technology in the medium term. He can plan to transition to a Database Administrator. He can identify the tasks where he has no experience or skills in. He can then make plans to either gain some experience in those tasks (by asking for opportunities with his current employer), or attend skills and technology training in them (on his own, through his employer, or through government sponsored initiatives), or both.

When his current job is finally disrupted and displaced, he is ready to move on. Or he might already have moved on, in pursuit of his interests. Either way, he has become more resilient.

## Conclusion

Our recommendations prioritise helping people find meaningful work opportunities. We are aware of recent discussions about providing social benefits for those who are permanently unable to find new work. Those are important; they are also likely to be untenable if done on a large scale. Moreover, deciding now that we cannot do anything to create large scale opportunities for many is arguably premature. Focusing on tasks can create those opportunities.

# 4.3 Work in 2040

## Task Masters

*Work was transformed once again, as tasks steered skills and jobs.*

Once you can deconstruct into tasks, you can reconstruct into any combination and permutation. To be a master at tasks was to be a master of scale.

At the scale of specific tasks, it was easier to identify specific skills that students and workers needed to master tasks. It was easier to identify who could help them, either in the city or globally. It was also easier to identify and develop the specific technologies that could augment their performance of those tasks. Augmented with the right technologies, they could even transcend their age, abilities, and disabilities.

At the scale of individual citizens, master enough technologies, abilities, skills, and knowledge (TASK) for enough tasks, and each of them could be a master of deep craft.[76][77] They could become excellent or even the best in the world, whether they were still salaried, frequently freelancing, or somewhere in between. And that ensured they had a place in the sun.

It also helped workers displaced by automation and outsourcing. They were no longer limited to looking for new work that was like the jobs they lost. They could also scan sectors that valued similar tasks. This immediately expanded the opportunities available. It was also easier for

> To be a master at tasks was to be a master of scale: the dexterity to deconstruct into the scale of tasks gifted the capacity to reconstruct into the scale of the world.

prospective employers to see the fit, even if the workers were from a different sector.

For citizens then, being task-oriented meant they thrived more and struggled less.

At the scale of companies, mastering tasks meant mastering change. They could track which tasks in their companies were being disrupted by the accelerating advances in technologies. They could pinpoint which parts of their businesses and which employees were under threat, and thus respond swiftly. They could also watch which tasks in their local, regional, and global markets were being digitally disrupted. They could thus innovate, seize the opportunity to disrupt competitors, and grow their companies.

Companies could also strategise and organise work in new ways. They were no longer constrained by the expertise they could find around them. They could now access affordable expertise and technologies from both established and unconventional sources, and from anywhere in the world. Any company that wanted to thrive could now build an integrated local- and global-scale network to boost its competitive advantage.

It was the same for cities. For them, mastering tasks was a new strategy to do better and to do different. How to educate students. How to train workers. How to draw on the world. And how to draw the world to the city.

The dexterity to deconstruct into the scale of tasks gifted the capacity to reconstruct into the scale of the world.

# Artefact from the Future: Task Transition Framework

**Kim,** late 40s

Task Transition

Expert

## Year 2020: Task Transition Framework[78–95]

*Helping workers make regular and rapid transitions to new and more valuable tasks in times of technological disruption.*

Everyone agreed Kim was super sassy. She stood up for herself and others, with wit to boot. She cared about the company and her colleagues. She understood technology and what it could do. So it was no surprise that when the crunch came, she was entrusted with the task.

It was an important task. She was to help her colleagues quickly transition to new tasks within the company. If only she had more time, she thought.

The new company management could not wait to use technology to automate as much of the company's work as possible. The old management had been slow to capitalise on artificial intelligence, robotics and Big Data. It was time to rectify that. The sooner the better. The company would be a lean, mean, money-making high tech machine. The new management gave Kim three months to determine what could be automated, and which employees would stay or go as a result.

Kim set to work immediately.

She felt that the right "resolution" was to focus on the tasks that people did within their jobs and not on jobs per se. She wanted to hear directly from those who actually did the work day-to-day. She spent hours each day talking to as many of her colleagues as possible.

She did all this face-to-face. She knew she could have saved some time if she messaged, multi-media emailed, or multi-reality-sensory mobile-conferenced them, but she also sensed face-to-face was best. Because she was genuine and jaunty, her colleagues warmed to her, even as they knew their livelihoods were at stake.

The most heartfelt sentiments came from her colleagues who were in their 60s. They had experienced the business process re-engineering movement of the 1990s. They remembered that the movement's use of information technology and the focus on the outcomes of tasks had achieved process and productivity improvements. They also rued that in many cases it had alienated, retrenched, and disillusioned people. They confided in her that they had a foreboding sense of déjà vu. They advised her to look carefully into the potential, perils, and pitfalls.

Her technical colleagues pointed her to Agile Development and Scrum. Developed and used since the 1980s, its methods include breaking down entire projects into tasks in such a way that large teams could work on them collaboratively and productively. It looked like a successful process, one that had many useful lessons to offer.

Kim herself was concerned about "Digital Taylorism". When she was in school, she had read about the promise and problems of early 20th century scientific management. She had even read Aldous Huxley's "Brave New World" and watched Charlie Chaplin's "Modern Times", satires of scientific management's dehumanising effects. She knew that even though she could now use technology to accurately measure what each of her colleagues were doing, it did not mean that she necessarily could or should measure everything. She wanted to be sure she did not end up dehumanising the company.

She integrated all that she learnt into a framework: the Task Transition Framework. She could use this framework to determine what and how tasks could be automated, and which of her colleagues would stay or go as a result (see Fig 4.3.1).

This was how the Task Transition Framework worked:

## a) Step 1: Break down what everyone does into tasks.
Companies have done this for much of modern industrial history. Be it scientific management from the early 1900s, total quality manufacturing, re-engineering, Six Sigma movements in the second half of 20th century, and the extensive use of Customer Journeys in design and Agile Development in product development.

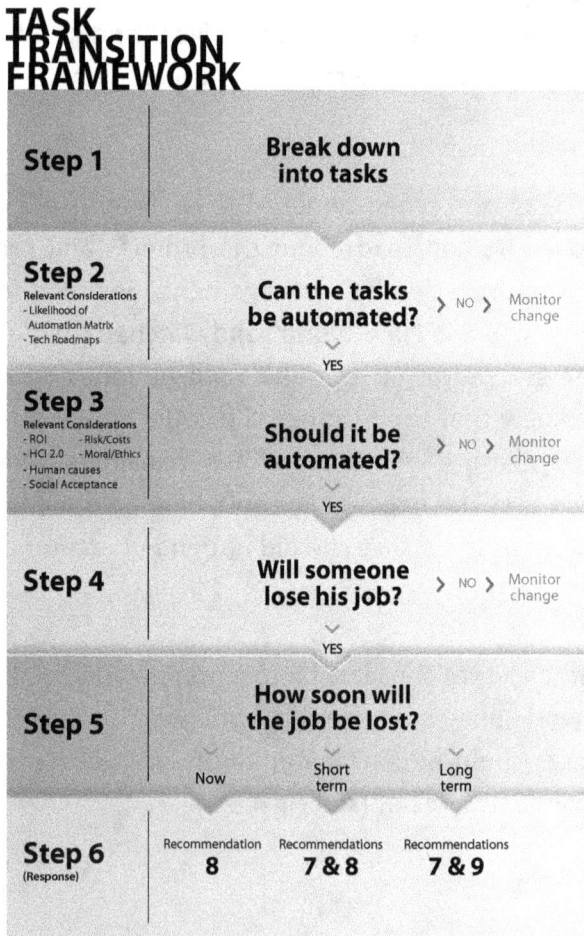

Fig 4.3.1 — Task Transition Framework

## b) Step 2: Assess if the tasks can be automated based on the matrix below (augmented with examples from healthcare):

| | Tasks that are:<br>1) Variable but predictable<br>2) Routine<br>3) Codifiable | Tasks that are:<br>1) Variable but unpredictable<br>2) Non-Routine<br>3) Uncodifiable i.e. needs tacit knowledge, flexibility, and judgement |
|---|---|---|
| **Tasks requiring limited sensory-motor coordination** | **Likely in the short term** by automating with existing/emerging tech<br><br>e.g. turning patients over in their hospital beds | **Possible** if technologies and tasks are re-designed<br><br>e.g. positioning patients correctly in an operating theatre |
| **Tasks requiring fine sensory motor-coordination** | **Likely over the long term** with accelerating advances in tech<br>e.g. stapling of wounds for some surgeries | **Maybe** if technologies and tasks are re-designed<br><br>e.g. sewing up wounds through the DaVinci machine |
| **Tasks typically considered as knowledge work** | **Likely over the short term** especially those with formal logical rules<br><br>e.g. monitor dementia patients for erratic but typical behaviours at night; dispensing medication (even as medications are more complex); basic health screening | **Unlikely** unless technologies pass or come close to passing the Turing test<br><br>e.g. reassuring patients; encouraging students; dealing with pandemics |

Fig 4.3.2 — Likelihood of Automation

*The rows and columns in the above table were sparked by the discussions found in* Why are there still so many jobs? The history and future of workplace automation,[93] *and built upon through subsequent discussions, workshops and further research (for example, our project comprehensively reviewed the trends in surgical robots as they are considered the state-of-art in the evolution of robots; our review gave us a good idea of how robotics will make steady progress in degrees of freedom, multi-functional capabilities, autonomy, tolerance for human error, reduced training needs, and lower operating costs.)*

In brief, tasks that are routine, predictable, and codifiable will likely be automated in the short to medium term. This will be the case whether the tasks are physical in nature, or cognitive/knowledge-based in nature.

Tasks that are non-routine and unpredictable, needing flexibility and judgement will take longer to automate. It might be possible to automate some of them. Automation for the rest will remain unlikely.

Technology trends should also be assessed for game changing innovations that could overcome current limitations that prevent technologies from taking over a human's job. These technological limitations include:

| Technological Limitations | Examples |
|---|---|
| Perception and manipulation | Finger dexterity; Manual dexterity; Cramped work spaces needing awkward positions |
| Creative intelligence | Originality; Fine arts |
| Social intelligence | Social perceptiveness; Negotiation; Persuasion; Assisting/caring for others |

Fig 4.3.3 — Technological Limitations
*Summarised from Frey, C. B., & Osborne, M. A. (2013). The future of employment: how susceptible are jobs to computerisation.*[95]

**c) Step 3: For the tasks that can be automated, decide if the tasks should be automated:**

   a. Consider the ROI (i.e. Return on Investment)

   b. Consider the risks and costs of the automation, especially those of an error arising from the automation (e.g., false negatives or positives; if there is an accident with the machine)

   c. Consider if there are moral and ethical reasons not to do it (e.g., discriminatory).

   d. Consider social readiness/acceptance

   e. Consider human causes (see Drivers of Change)

   f. Consider if long term human cognitive abilities to develop expertise will be compromised (see HCI 2.0 in the box below)

---

**Premise of HCI 2.0 (Human-Computer Interaction 2.0)**

"HCI 2.0" is an overarching theme we coined to account for several emerging trends across several fields that we observed in our research for *Living Digital 2040*.

When we begin to rely on technologies for substantial cognitive tasks such as identifying patterns and making decisions for us, we are diminishing the intensity of our thinking. How many times have we felt a suspicion that our arithmetic ability is getting worse because we rely on our calculator app all the time? How often have we noticed that we don't seem to remember our friends' birthdays and anniversaries because now we are conveniently reminded by our calendar app or birthday-reminders.com?

Software tends to undermine users' "ability to encode information in memory, which makes them less likely to develop the rich tacit knowledge essential to true expertise".[96–98] Nobody is certain of the exact mechanisms or the extent to which this will affect our cognitive abilities as it is only recently that the penetration of technologies into our everyday lives became so deep.

There are already many anecdotal and qualitative studies that show the danger. A series of experiments reported in Science (2011) indicated that the ready availability of information online weakened our memory for facts.[99] Simply knowing that an experience has been photographed with a digital camera weakened a person's memory of the experience.[100]

When we start relying on automation for conducting our cognitive tasks such as pattern recognition and decision making, we "hamper the mind's ability to translate information into knowledge and knowledge into know-how",[101–103] and evidence can be found in many recent literature, e.g., in game playing ,[104–106] in accounting,[107] in financial trading,[108] in programming[109] and in way-finding.[110][111] Reliance on GPS navigation tools over many years reduces hippocampus function in our brain, resulting in possible likelihood of developing dementia.[112]

Again an analogy can be made with the effects of diet or lifestyle in a person's life. There is a difficulty of clearly establishing cause and effect when it takes many years and decades to study empirically.

We need to pioneer a new field, HCI 2.0, to examine these longitudinally. Celebrating our increased efficiency, productivity and convenience is good, but we need to start thinking of designing our technologies to support these 'tangible' dimensions without compromising, or at least with clear awareness of, the long-term impact it has on our cognitive abilities.

**d) Steps 4-6: Determine who will lose their jobs, how soon, and how to help each of them (see Recommendations 5, 6, 7, and 8).**

Kim was ready to try out the Task Transition Framework. She had been given three months to give her recommendations, and the days and weeks were flying by. Her colleagues were getting visibly more anxious. They would come by to speak with her, ostensibly on work, but what they really wanted to know was if they should prepare for life outside of the company.

Kim knew that being sassy and witty had helped her so far in managing her colleagues. To pull all this off though, she would need all the wits that she could keep about her.

# Artefact from the Future: Redesigning Recycling 2020–2040

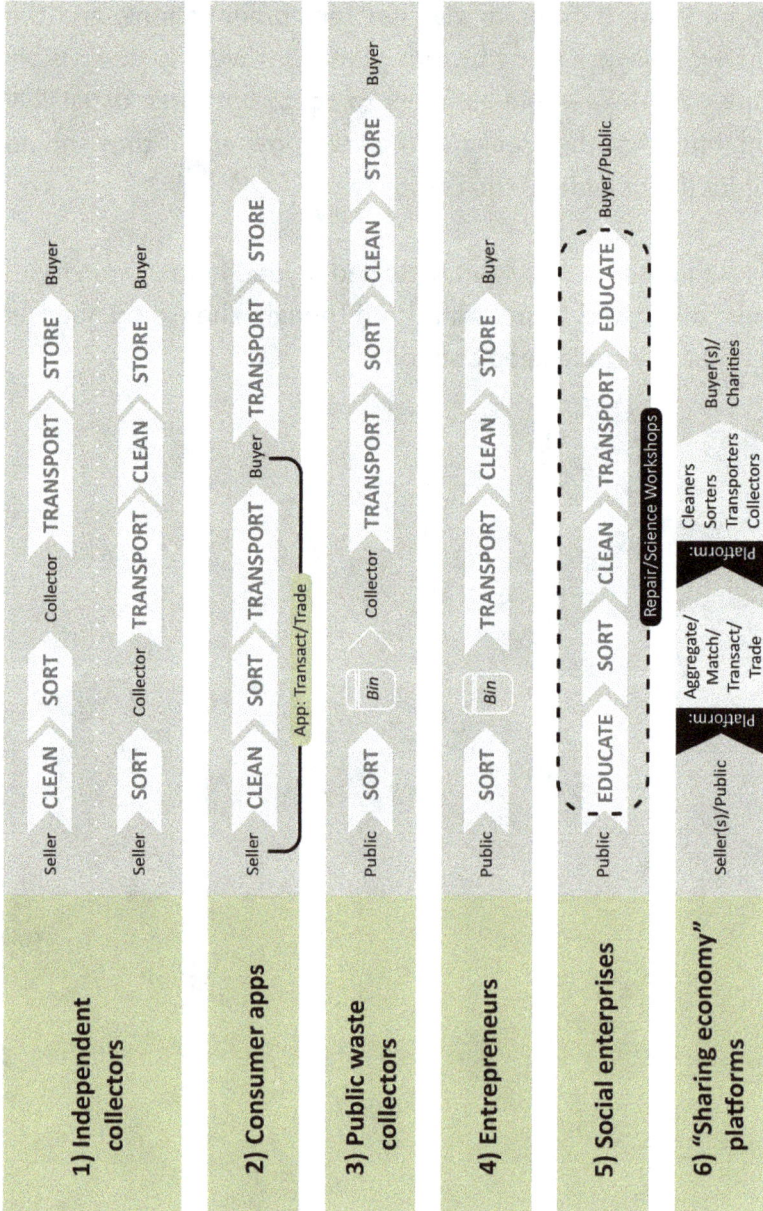

**1) Independent collectors**

Seller › CLEAN › SORT › Collector › TRANSPORT › STORE › Buyer

Seller › SORT › Collector › TRANSPORT › CLEAN › STORE › Buyer

**2) Consumer apps**

Seller › CLEAN › SORT › TRANSPORT › Buyer › TRANSPORT › STORE

App: Transact/Trade

**3) Public waste collectors**

Public › SORT › Bin › Collector › TRANSPORT › SORT › CLEAN › STORE › Buyer

**4) Entrepreneurs**

Public › SORT › Bin › TRANSPORT › CLEAN › STORE › Buyer

**5) Social enterprises**

Public › EDUCATE › SORT › CLEAN › TRANSPORT › EDUCATE › Buyer/Public

Repair/Science Workshops

**6) "Sharing economy" platforms**

Seller(s)/Public › Platform: Aggregate/Match/Transact/Trade › Platform: Cleaners Sorters Transporters Collectors › Buyer(s)/Charities

Fig 4.3.4 — Redesigning Recycling and the Circular Economy

Notes:

1) Traditionally, independent waste collectors went from door-to-door (alerting households with a characteristic honk). Known colloquially in Singapore as "karang guni" or "rag and bone man", they would pay for and pick up unwanted items. The households disposing of these would likely have done a rudimentary sorting of their items. Some would have cleaned them beforehand, but not always. The independent waste collectors would then cart these items away to be sold, earning an income for themselves.

2) With the advent of digital technologies such as apps, sellers can now list items online for buyers to pick and purchase. The seller and buyer would agree to make their own way to somewhere mutually convenient for the exchange of items and monies. An example of this is the Singapore-based app Carousell.

3) Members of the public dispose of their unwanted items into bins, sorting different materials (e.g., paper, plastic, glass) into different bins. No payment is received. Public waste collectors would then pick up the items from the recycling bins. Unfortunately during the transportation process, the items can become mixed up i.e., undoing the early sorting. The public waste collectors would then re-sort and clean the items to be stored and sold subsequently.

4) The above immediately creates an opportunity for an enterprising firm to provide a transportation service that does not inadvertently mix up the different items. This service could be complemented with a bin re-design that encourages members of the public to dispose of the right items in the right bins.

5) Another opportunity is for social enterprises to redefine re-cycling as an opportunity for education and training in repair/repurposing. Members of the public and even students can be en-couraged to bring their unwanted items to community centres, schools, supermarkets, and other communal points or even lock-ers for repair and/or maker movement workshops. They can then reuse/purpose these items or sell them.

6) Opportunities are greatly expanded with the use of digital technologies. Digital solutions can be created to aggregate items from sellers and members of the public, match items of high and low value, and consolidate demand from potential buyers or par-ties in need, such as charities. There might also be opportunities to create a task marketplace for freelancers or firms to provide ser-vices in cleaning, sorting, transporting, and collecting. The larger opportunity then becomes one of how to configure all of these intelligently.

# 5  Future of Education

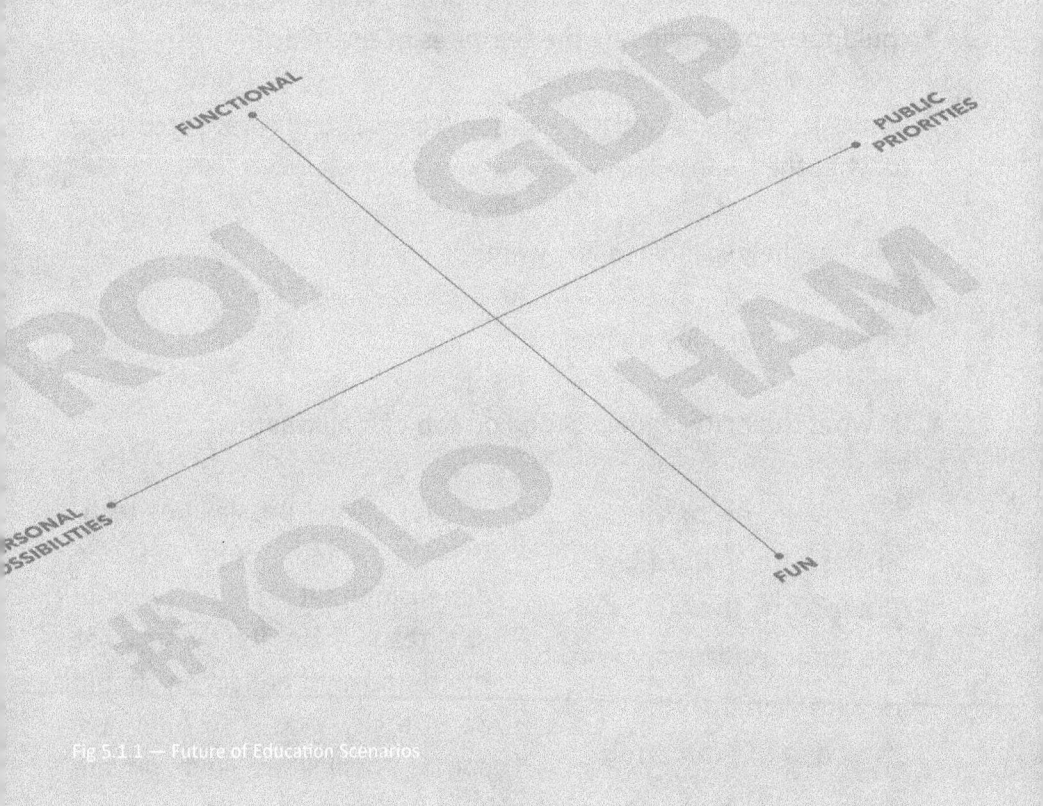

FUNCTIONAL

PUBLIC
PRIORITIES

PERSONAL
POSSIBILITIES

FUN

Fig 5.1.1 — Future of Education Scenarios

## 5.1 Scenarios[1–46]

It was a world of possibilities. Technology had turned education into a web of choices.

Lessons could be self-paced or peer-tutored. It could consist of real-reality learning journeys or multi-reality learning experiences. It could be inside-out, starting with theories and concepts, or outside-in, starting with problems and challenges. It could be done in person, one-on-one or in a classroom; or, there was an app for that, as well as a teach-bot or MIDAs (Massive Interactive Digital Academies). It could be roll-up-your-sleeves and hands-on, or it could focus on expanding the "empires of the mind".

Students, teachers, and parents could choose and curate according to what they wanted.

Or what their school leaders wanted.

Or what companies wanted.

Or what their city leaders sitting on top of it all wanted.

> **Yet, for all the possibilities, there were two dimensions that technology could not resolve on its own: the purpose and objective of education.**

Yet, for all the possibilities technology created, there was one dimension that technology could not resolve on its own: what the purpose of education should be in a city. Was it to fulfil personal possibilities and private aspirations? Or was it to meet public priorities and achieve societal goals?

That dimension gave rise to a related one. Was the objective of education functional, geared towards helping school leavers find jobs, grow economies, and improve societies? Or was the objective of education fun, oriented to help students discover their interest, passion, and *raison d'être*?

These were decisions cities, companies, and, citizens had to make.

Whether it was driven from the ground by citizen choices, negotiated among citizens, companies, and cities, or imposed from the top by cities, their decisions along the "personal possibilities-public priorities" and "fun-functional" dimensions defined the future of education.

## GDP

Those who chose "GDP" decided that the purpose of school was to train workers to meet national economic (and some social) goals.

It was inevitable. Worldwide growth had lost speed, as the growth in many emerging economies slowed due to the pace of structural and regulatory reforms. As growth slowed, cities and countries competed even more aggressively. They all thought there was not enough to go round for everyone. Better to grab what there was.

Many of them relied on the one strategy that had worked in the past: building work-ready capabilities. Technologies had made it possible to educate and train more people more quickly. If a sufficiently deep pool of industry capabilities could be built, it would attract investments. The investments would drive economic growth and the creation of indigenous startups. These in turn would provide the resources to further build and develop cities and citizens.

Some criticised this bend-to-the-economy strategy. But these critics grudgingly acknowledged the strategy had helped to meet the demands on the public coffers. And the demands had grown tremendously. Aging populations needed to be cared for, aging infrastructure had to be maintained, and smart cities required huge investments. And someone had to pay for all the technologies that made cutting-edge education, training, and upgrading possible.

The students whose interests were aligned with the public priorities were the lucky ones. They studied what they wanted, and not what they had to.

Parents and teachers did not all agree with the system, but they also saw that it had opened wide doors for careers. They remained concerned for those whose interests did not fall in line with the public priorities. The teachers tried their best to make the classes "relevant" and interesting. There was some success, but it was not across the board. Parents began contemplating sending their children overseas on alternative education pathways and work trajectories. For many of these parents, they privately felt thankful that at least the economy was steady, providing them with the work and incomes that made sending their kids overseas an option they could even contemplate.

Fig 5.1.3 — Future of Education Scenario - ROI

## ROI

While some cities, companies, and citizens chose "GDP", others chose the "ROI" route. Those who chose "ROI" decided that the purpose of school was to train workers so that they could secure good jobs for themselves.

"ROI" was a close cousin of "GDP": it too was about being able to compete locally and globally. But it was the choice of cities, companies, and citizens who felt "GDP" smacked too much of industrial policy. Pointing to the past, they were sceptical that governments could pick the right winners. Better to let the capabilities emerge organically, and the best way to do that was for citizens and companies to lead the way.

Inevitably, parents and students chose areas of study that had a high payoff. Very often, this was in the form of high pay. If they could not get high salaries immediately after they left school, then at least there should be some prospect of higher salaries in the medium term.

Schools did not have to worry about making lessons interesting to students. But that did not mean the schools had an easy time. Lessons did not need to be interesting; but they had to be useful. Parents and students often challenged principals and teachers on why they had to learn a certain subject.

Even worse (or better, depending on perspective), because technology had made it possible to slice and dice lessons into nano-content, parents and students would often challenge the schools on these too. It was a never-ending tussle. Stressed out teachers told of waking up in cold sweat, after nightmares of parents and students screaming "what jobs will this be used in?"

There were critics of course. There were those who believed that education was not just about getting a job. They were worried about the "vocationalisation" of education. Others felt that education should be broad-based. Being able to draw on knowledge across disciplines was fundamental to an innovation economy.

Another group argued this favoured families of higher socioeconomic status, who had more money, contacts, and time to invest in their children outside of formal schooling.

The upside was that students made a seamless transition into the workforce. Companies were pleased that school leavers had such strong and relevant academic basics. They did not mind paying higher salaries for them. Cities were pleased too. They believed in the mantra that if students and workers flourished, cities would flourish too. Of course, they often omitted that what they meant by flourish was largely measured by ROI.

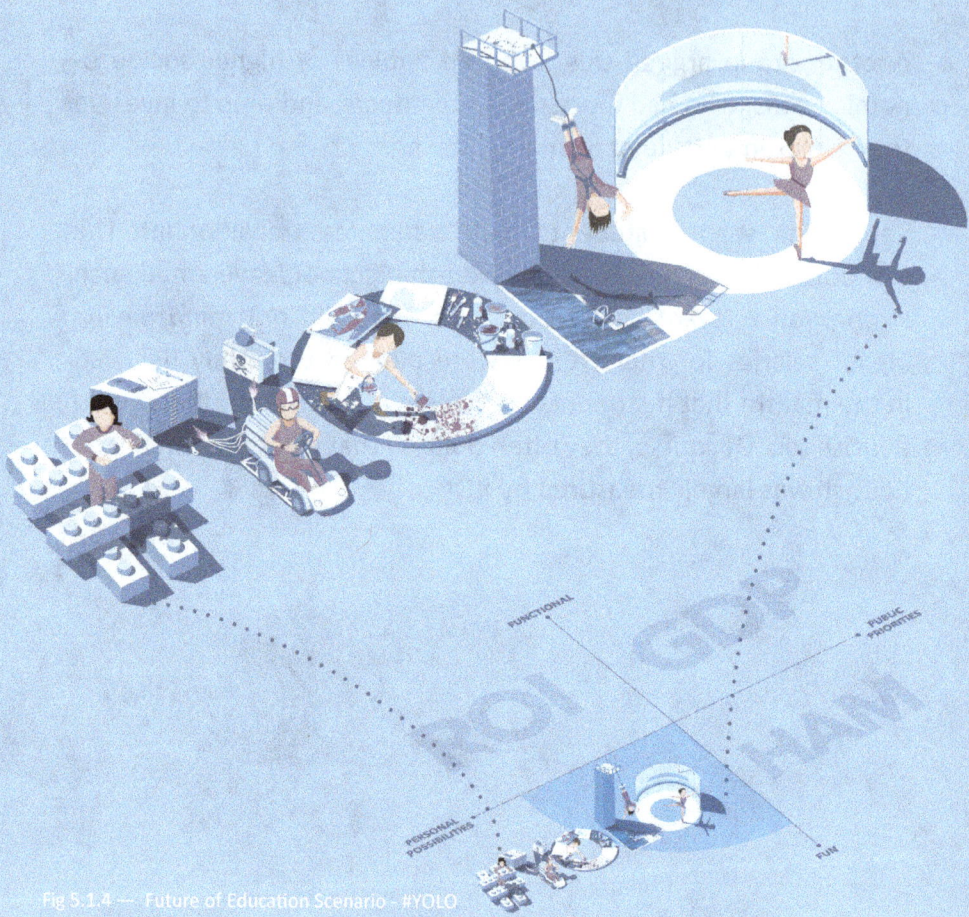

Fig 5.1.4 — Future of Education Scenario - #YOLO

# #YOLO

Diametrically opposed to "GDP" and "ROI", were the cities, companies, and citizens who believed education should be all about the pursuit of fun.

"#YOLO" believed in the idea that "function follows fun", managing to turn modernist design principles on their side (and their pioneers in their graves), and to lay out its educational philosophy at the same time. What mattered was play, and whether students enjoyed what they did. The utility will follow.

They cited studies from various countries showing how play was important for motivation and social development. They pointed to studies of genetics and pedagogy that found that formal study of certain subjects could start later, without any loss of academic achievement subsequently.[24] They emphasised that children and teenagers are only young once in their lives.

They admitted that this might not have been feasible in the past. But now, technology had made all this possible. When you can learn almost anything at any time, and when you can learn from anyone in the world, and when you can accelerate your learning with analytics, there is time to play and have fun first. [25-27]

The #YOLO parents and students demanded that cities structure their schools to follow this philosophy. Lessons had to be enjoyable, engaging, and interesting. No matter how obscure any student's interest was, the school was expected to cater to it (because technology had made it feasible).

Schooling was no longer the culprit that "took away the light in students' eyes". The eyes sparkled. Parents and teachers noticed students were more interested in their studies. They also noticed students were prepared to go the extra mile and were amazed at some of the creative ideas the students sometimes came up with. They felt it bode well for the future.

But to companies, it was not so clear the school leavers they hired were quite so sparkly. Company hiring managers wondered if the students had a strong foundation of the basics. They also saw that they had to provide more work-specific training. And their new hires were often very easily bored: making work fun was hard work.

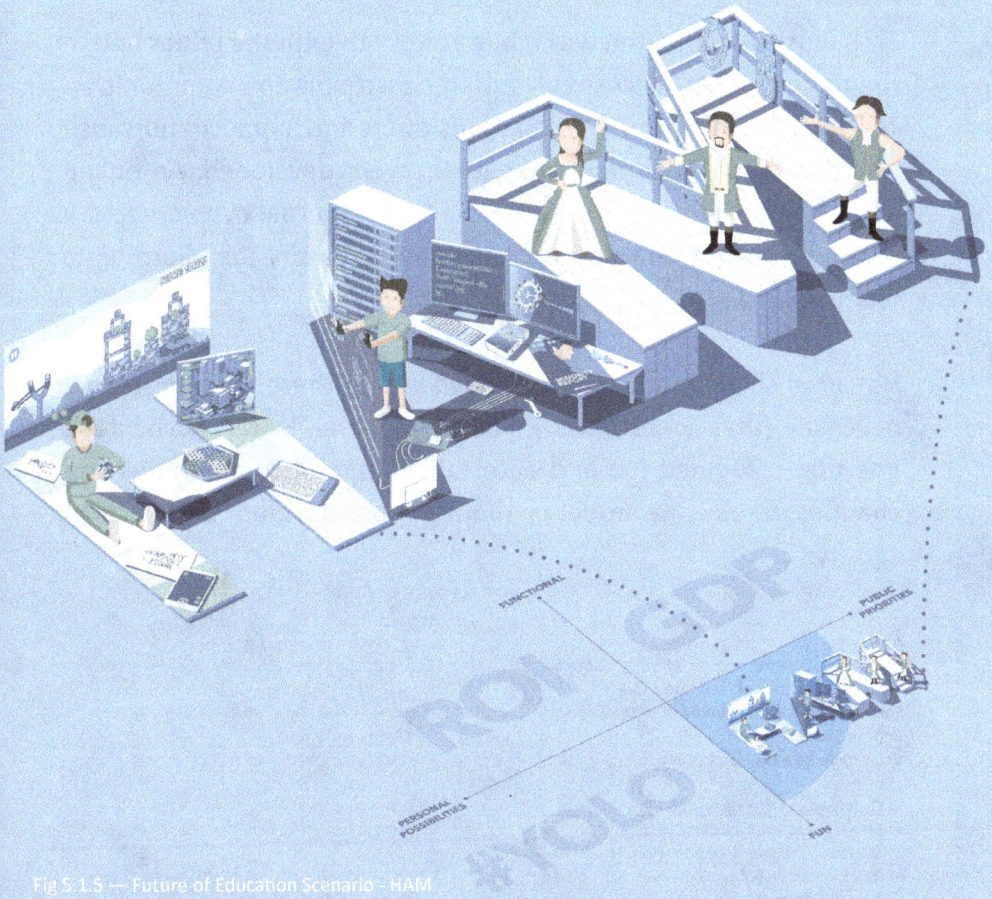

Fig 5.1.5 — Future of Education Scenario - HAM

# HAM

The "HAM" set of cities, companies, and citizens believed that through musicals and movies, education could be fun and still meet public priorities.

Hip hop. Smash hit. Class lessons. Who would have thought a Broadway hip-hop musical about a historical figure would make its way from stage to school, and inspire many around the world?

That musical, *Hamilton*, was a huge hit, both with the critics and at the box office. Its success galvanised classrooms and communities. Teachers taught with it. Foundations offered to support and subsidise tickets for students. Study guides were developed. An online education portal was built. The Wall Street Journal even created a computer algorithm to deconstruct and demystify the allure of its "complex rhyming lyrics" (see Fig 5.1.6).[30–36]

*Hamilton* showed educators worldwide that there was a potential new way (they were quickly nicknamed the "HAM" advocates): one that combined fun and entertainment, and communities and charities, to educate students about important values and ideas in society.

**A) Original lyrics from musical**
How does a bastard, orphan, son of a whore and a
Scotsman, dropped in the middle of a
Forgotten spot in the Caribbean by providence
Impoverished, in squalor
Grow up to be a hero and a scholar?

**B) Deconstruct lyrics using CMU Pronouncing Dictionary**
a) Break up each word into syllables
b) Map each syllable to phonetic sound
c) Check which syllables rhyme with each other
(e.g. vowel sounds, consonant sounds, and stresses)

**C) Pass through an algorithm**
a) Assign scores to the rhymes
b) Cluster rhymes based on scores
c) Improve visualisation – treat as similar to
travelling salesman problem in computer science

**D) Lyrics with rhymes clustered**
How does a bastard, orphan, son of a whore
and a Scotsman, dropped in the middle of a
Forgotten spot in the Caribbean by providence
Impoverished, in squalor
Grow up to be a hero and a scholar?

OR rhyme: or; whore
AH-N rhyme: phan; son; and; man; in; en ;
in; an; dence; Im; in
AA-T rhyme: Scots; dropped; got; spot
IH rhyme: midd; i
ER rhyme: For; Car; er; lor; lar
AA rhyme: pro; pov; squa; scho
OH rhyme: Grow; ro

Fig 5.1.6 — Integration of Hip Hop, Rhyme, Digital, and Education
*Adapted from WSJ (see reference). This also demonstrates the potential of digital technologies in school e.g., deconstruct popular songs.*[30–36]

"HAM" coincided with two other trends. The first was the increasing use of real science in science fiction movies. Science fiction had always inspired science. But since movies like *Minority Report* in the 2000s started paying more attention to scientific research, science was also inspiring science fiction.[37] And in the 2010s, movies from *The Martian* to *Ghostbusters* all endeavoured to be as scientifically accurate as possible.

The second was the use of games in the classroom. From flinging furious fowl at giggling piggies to learn science concepts, to catching pocket monsters to discover new places and people,[38–46] these built on longstanding efforts of adopting and adapting games for education.

All these meant that movies, musicals, and many other types of entertainment could be used to achieve serious educational goals. In a world short on attention but saturated with media, they were able to hold on to students' attention, sometimes for hours.

Technology helped in many ways. Firstly, it became much easier to find these sources of science in entertainment. Secondly, the sources could be curated into a class. Thirdly, the sources could be easily shared with other teachers and students. Techniques such as "gamification" also expanded the tools available to keep students motivated. Lastly, technology helped to keep costs down, increasing accessibility.

It really did feel like the "HAM" group was on to something. But it had its critics. After all, imbibing values through fun public education sounded too much like mass propaganda. Was that not what nation- or city-wide games, festivals, and parades aimed to do?

Other critics who did not think it quite warranted that label, pointed out that the costs would be exorbitant and prohibitive.

And new concerns arose: was this effectively abdicating the public role of schooling to commercial entertainment?

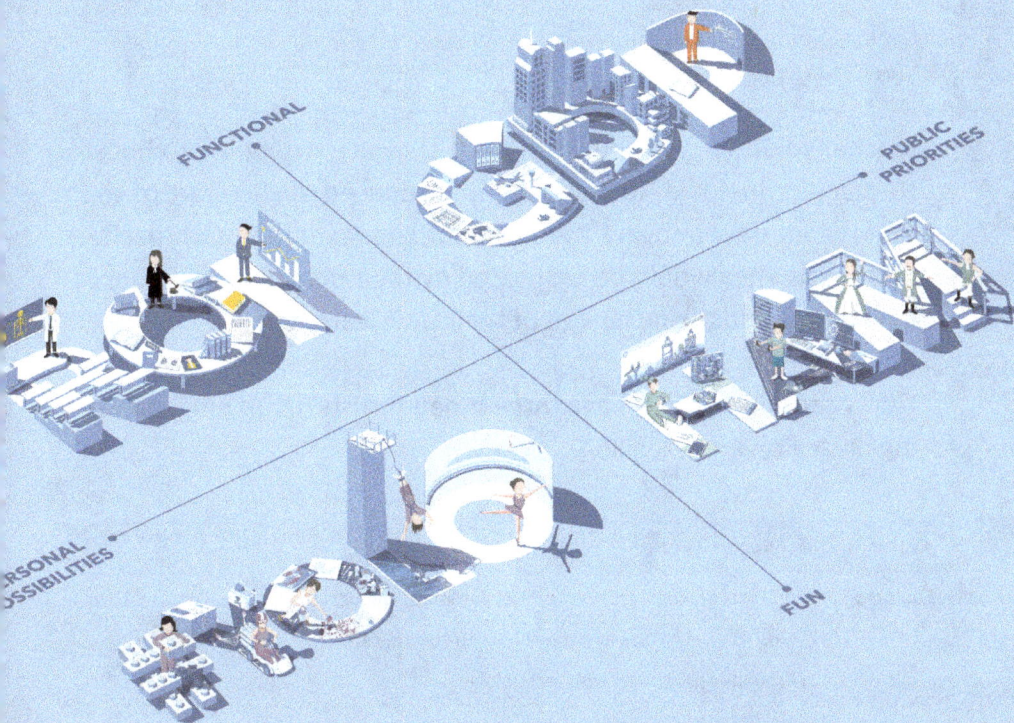

Fig 5.2.1 — Responding to Future of Education Scenarios

## 5.2 Recommendations

The future could pan out into one of the four scenarios. It could also be a mix of them. We want to be ready for all of those possibilities, which would give us the greatest agility to adapt.

If your city has a high quality education system, what else can you suggest to it? We can suggest ways to use digital to build on that strength. Digital means education can now scale from one student to the whole city (and even the world). We can reimagine how we help more students and teachers succeed on their terms, while closing digital divides and gaps. And of course, we can give ideas about how to raise the average performance of its students.

Our recommendations thus have **a unifying aspiration: personal peaks.** Digital opens many doors to nurture each student or teacher to attain their peaks in what they are strong in or choose to excel in. Just like Adi, whom we introduced at the start of this report, but now imagine the same for every student and teacher in the city. They would be motivated by their pursuit of excellence, taking risks and taking charge of their own learning.

Our recommendations have three broad thrusts: equipping, elevating (at scale), and excelling.

| Approach | Recommendations |
|---|---|
| Equipping | 1) Build technologies to modularise, automate and personalise teaching and learning. |
| | 2) Close digital divides and the "now-can" digital gap. |
| | 3) Expand experiments with technologies. |
| Elevating (at scale) | 4) Peer tutor digitally at city-scale to raise educational effectiveness. |
| | 5) Nurture empathy at the city-scale. |
| | 6) Peer tutor digitally at the global scale to seed global citizenship. |
| | 7) Help each student/teacher access an affordable global team of mentors. |
| Excelling | 8) Spot students' strengths and talents by erasing the line between school and work. |
| | 9) Be agnostic about measures of success. |

Fig 5.2.2 — Summary of Recommendations for Future of Education - Equipping, Elevating (at scale), and Excelling

# Equipping

**1) Build technologies — such as Lesson Design Maps and EduBang — to modularise, automate, and personalise teaching and learning**

The astute teacher will look at the scenarios and ask a very practical question: how do I prepare and teach my lessons? The teacher of tomorrow will have to make the lessons interesting and fun, but at the same time useful and functional. He will also have to personalise the lessons while serving a larger public purpose.

We have to help teachers to do this as effortlessly as possible. Lesson Design Map and EduBang — two Artefacts from the Future that we created for *Living Digital 2040* — illustrate what we mean.

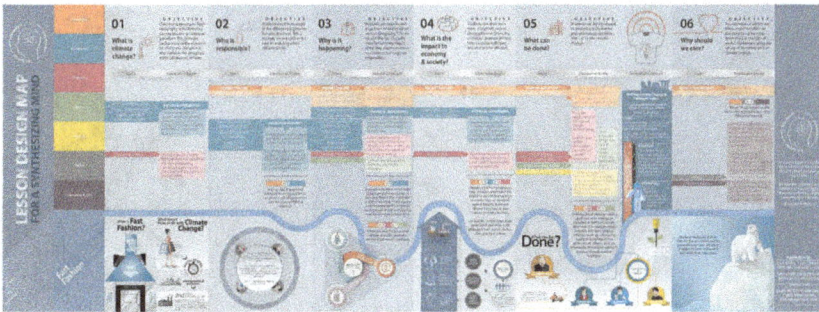

Fig 5.2.3 — Lesson Design Map (see also Artefact from the Future in section 5.3)[47]

Lesson Design Map (see section 5.3 — Artefact from the Future) modularises: it takes different subject syllabi and breaks them down into modular concepts and topics. These modular pieces are then rebuilt into connections between subjects.[48-50] They are also rebuilt into connections across different scales: from the global (such as climate change challenges), to the personal and everyday (such as fast fashion). Tying this to the formal curriculum is key. The different subject syllabi have been rigorously put together. Building on that strength helps schools do more, more quickly.

While Lesson Design Map modularises, EduBang automates (see section 5.3 — Artefact from the Future).[51] Increasingly accurate computational technologies such as computer vision and natural language processing enable meaningful modular materials (e.g., videos, movies, music etc.) to be composed and combined into new materials. They will be easy to search, browse, create, and disseminate. With so much content now available, technologies like EduBang can do this not just for any syllabus but for any content with embedded educational value. They can also automate the reassembly into different and larger lessons.

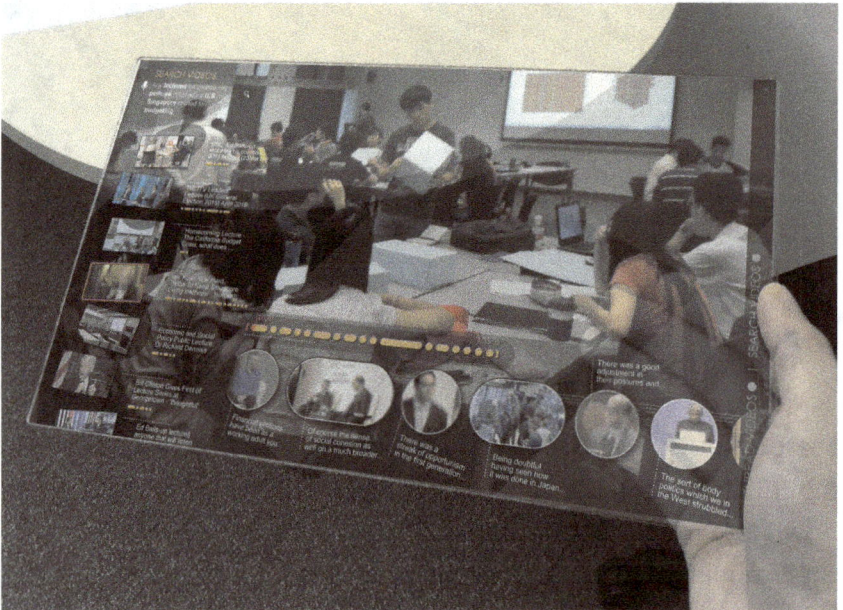

Fig 5.2.4 — EduBang (see also Artefact from the Future in section 5.3)

EduBang makes it easier to use the Lesson Design Map. Teaching with the Lesson Design Map is best executed by a team, but that makes it more challenging to implement. With technologies like EduBang, teachers of different subjects could easily assemble relevant content for each other. This content could in turn be assigned to students for self-paced learning, flipped classrooms, or peer tutoring. Students could also assemble the content for themselves as an assignment or project.

**Pedagogies and models will continue to shift and transform. Technologies like Lesson Design Map and EduBang mean teachers can assemble lessons quickly according to how they teach and how each student learns. Schools can not only personalise learning; they can also personalise teaching.**

Once we can deconstruct and reconstruct this way, the possibilities expand rapidly. From the Gutenberg printing press, to car manufacturing (Ford in the 1910s, Toyota in the 1950s), to modular IT platforms, modular and interchangeable parts have always transformed social and economic sectors.

Moreover, in the next two decades, the "right" pedagogies and models will shift or be transformed. Technologies like Lesson Design Map and EduBang can be designed to be "agnostic" — teachers can assemble lessons quickly according to any pedagogy, teaching modality, and learning style. They can also assemble them according to how they teach and how each student learns.

A lot of the focus so far has been on personalised learning. Now schools can also personalise teaching.

## 2) Close the "now-can" digital gap: raise parents' digital literacies and close digital divides

Lesson Design Map and EduBang demonstrate one thing: what we do now with digital often lags behind what we can do with it. This "now-can" gap could widen as technological advances accelerate.

Schools often help students and teachers raise their digital literacies[50–52] to close this gap. This is already done well through the systematic introduction of relevant technologies and training. One teacher — not known to mince her words — summed it up best when she shared that the Singapore Education Ministry was "good at providing resources, infrastructure and training... and there are conscientious people thinking about the future."

What happens in school can be easily reinforced or unravelled at home (and vice-versa). Raising digital literacies across teachers, students, and parents (especially those from less privileged circumstances) ensures we can open up opportunities to all.

We can do more. If we envision parents as equal partners in education, and part of a larger learning community, then raising digital literacies is important for parents too, and not just for teachers and students. For those parents who are interested, it is time for schools to help parents "go back to school". This can be done through online courses, digital peer tutoring[52–59] (see Recommendation 4), or face-to-face classes conducted in school, in community centres or by external providers. The course content will have to be aligned to what is happening in their children's schools. Otherwise, if there is a digital gap between parents and teachers and students, it will constrain how seamlessly they can all work together.

This is especially so for parents and students from less privileged circumstances. Educators we spoke to stressed that the nurturing and education in school can be easily reinforced or unravelled at home (and vice-versa). Many parents want to play a more active part but are constrained by their circumstances. Raising their digital literacies will ensure they are not inadvertently left behind by the march of technology, and help narrow some of the gaps in out-of-school support for students from lower socioeconomic backgrounds.[60]

In summary, closing the "now-can" gap means we can now open up opportunities to all.

### 3) Expand experiments with technologies: be creative in making digital work for us

Schools are already experimenting with "flipped" classrooms, blended learning, games and gamification, real world projects, adaptive testing and learning, and education ICT of all stripes.[61][62]

Raising digital literacies makes it possible for schools to expand experimenting with new technological tools in creative ways. With the push for greater usability and more intuitive interaction modalities (such as touch, gesture, speech, and their combinations), especially in the last decade, the number of such tools have proliferated.

This can be simple, such as curating YouTube videos to prepare for science exams and Olympiads. The latter is a true story a teacher told us. The student had initially felt too sheepish to admit it, but when he finally did, the teacher told him that she thought he was absolutely creative and brilliant to have done that. (And as a further example, albeit in healthcare, when faced with an unfamiliar surgery, a nurse told us she prepares by reading online journals and watching YouTube videos to "understand the surgery process", so that she knows "what to expect from the surgeons").

These can also take advantage of emerging capabilities in data analytics and artificial intelligence. For example, the online tool mentioned in the HAM scenario could be used for poetry, rap, rhymes, and algorithms. Another example: there are learning analytics solutions which can predict within the first two to three lessons whether a student will struggle for the rest of the semester. The professors who shared their experiences with us told us that they use them to intervene early, so that they can put the student back on track quickly instead of waiting till crunch time just before tests and exams.

Schools will also have to consider how best to integrate technologies – from robotics to AI – to augment teachers. We will need to learn to design these technologies to work seamlessly with teachers and students. For example, Stanford University's report *Artificial Intelligence and Life in 2030* states that "[r]esolving how to best integrate human interaction and face-to-face learning with promising AI technologies remains a key challenge."[63]

Expanding experiments must come with clear ideas of how to increase their odds of success. Besides ensuring top-down leadership support, the technologies must also show tangible benefits as to how they are

Stanford University concluded that "[r]esolving how to best integrate human interaction and face-to-face learning with promising AI technologies remains a key challenge."

working for us. This has become even more important because, paradoxically, constant technology change can cause fatigue of and even resistance to new technologies.

# Elevating (at scale)

### 4) Peer tutor digitally at the city scale: increase educational effectiveness

Equipping students and teachers (and in some cases parents) with an expanded set of digital literacies and tools is only the start. Subsequently, we need to ensure we raise each student's educational performance.

According to the Education Endowment Foundation, the top two education strategies in cost and effectiveness are "feedback to pupils" and "meta-cognitive strategies (help students think about their own learning more explicitly)".[64] One solution would be to increase the time teachers work on these with students, or even to increase the number of teachers.

This is obviously not scalable, not to mention potentially prohibitively expensive. Digital technologies can help,[65] such as with learning analytics and adaptive testing. In future, AI engines (perhaps through chatbots and avatars) can augment this further. But as a teacher shared, the "interactions between teachers and students are the most magical, not the transference of knowledge". The human interactions still matter.

The scalable solution to this might very well be to combine digital with what the Education Endowment Foundation found as the third most effective education strategy: peer tutoring.

What if every year that they are in school, each student had to teach someone in the city as part of their formal education experience and development? They could do this as part of their community service or service learning. They could also do this as their "flipped classroom" graded homework. Digital makes this easy to do at the city-scale, and it can also be made more effective by augmenting it with learning analytics and artificial intelligence.

Properly structured and guided (e.g., with supervision from teachers and parents), the one teaching and the one being taught could reap mutual benefits. The one teaching would reinforce his own understanding of subjects. The one being taught will gain from the one-on-one feedback and guidance. A highly cited piece of research by Professor Benjamin Bloom — albeit conducted before we had so many digital tools — suggests that "an average student who receives one-to-one tuition will tend to outperform 98 percent of ordinary students in a classroom."[66]

Good ideas for teaching and learning will also spread more readily between schools, potentially lifting educational effectiveness across the city.

### 5) Nurture empathy at the city-scale: serve the public interests
Digital peer tutoring could have an additional benefit: we could nurture students' empathy towards others in society.

With digital pair- or peer-matching tools, students can be matched with a spectrum of peers across the city over their years in school. The peer can be a student who is from a less privileged background. It can also be someone who is not, and is just weak in a particular subject. It can be someone with a learning disability. Or just someone who wants to be energised about a particular topic. It can be someone of the same age. Or any age, from a younger child interested to learn more, to a curious adult who wants to pick up new knowledge.

Each peer pairing can be designed to achieve a certain outcome over a defined period of time so that it is not a once-off but a significant commitment. That way, the students teaching will need to understand the needs of their assigned peers, and would also be exposed to a broad spectrum of society and their different needs. They would thus develop a greater sense of empathy and shared responsibility for each other.

**The scalable solution might very well be to combine digital with peer tutoring. Digital peer tutoring at the city- and global-scale can help spread good ideas for teaching and learning, nurture empathy, and seed global citizenship.**

This is important. A principal encouraged us to think about where public interests fit in as learning becomes more individualised to students' interests. Encouraging a city-wide digital peer tutoring effort is a good way to balance this. Even as digital technologies can be used to serve the unique needs of one, they can also be used to serve the unique needs of many.

There is more. An interviewee, who is deeply passionate about healthcare and teaching, stressed how important it was to try to get people to know each other as humans. Empathy ensues once students realise the other person has similar traits, dreams or choices as them. He felt this was fundamental to building trust between people and in society.

Empathy might not be the only thing we can nurture across the city. We might be able to build trust too.

### 6) Peer-tutor digitally at the global scale: become a global citizen from young

If we can "scale" digital peer tutoring, why stop at the borders and boundaries of the country or city?

We can go global. We can encourage students and teachers to peer tutor anyone in the world. Again, with digital pair- or peer-matching algorithms and tracking tools, students can be matched with a broad spectrum of international peers. Their peers can be from any geography and age group, whether it is a high schooler from Asia, or a parent in America with two primary school children. They develop empathy for an international peer, and also a greater sensitivity to different cultures. These are the seeds of global citizenship and we can sow them from a very young age.

This also prepares them for a world that is globalised and continues to become more so. For small countries and cities, it is even more important. Through school, their citizens can become comfortable interacting across cultures and geographies from young. It is prohibitively expensive, even impossible, to send every student overseas for international exposure. But with digital peer tutoring, we can give every student a good grounding in global citizenship.

**7) Help each student and teacher access an affordable global team of mentors to develop her strengths: help each and every student or teacher peak (and not just in academics)**

We can teach the city and the world. The city and the world can also teach us.

As our starting story of Adi demonstrates, there might be times when the best person to help one of our students or teachers is not found in the school system. He or she could be in the city, country or even overseas, such as experts who are available for hire in the emerging global freelancing market.[67–69]

**The disruption, deconstruction, diversification, and democratization of expertise means we can affordably hire — especially for smaller cities — a global team of mentors for each student and teacher, who can thus become very good or even the best in what they choose to pursue.**

Trends suggest that hiring such experts need not cost an arm and a leg. Because the experts could come from any corner of the world, because we might be hiring them for only part of their time, and because they are helping us through the digital realm, they are becoming an affordable option. In fact, some might even volunteer to do it, because of the goodwill we have generated from teaching the community and the world (see Recommendations 5 and 6).

It is also easier to do this for small cities and countries. For example, Singapore has a yearly student cohort of thirty to forty thousand, making it easier to find experts worldwide for each of their students than for a country with a yearly student cohort of hundreds of thousands or even millions. The same goes for teacher numbers.

Imagine assembling, for each of our students and teachers, their own unique team of global experts and mentors to help them do better in both academic and non-academic pursuits. It will no longer matter if each student or teacher has few or many needs — digital global access will take care of that.[70-74]

Students and teachers could come closer to achieving their peaks. They could become very good or even the best in the world in what they choose to pursue.

local . regional . global

**Established Experts**

**1** Experts (doing it the way it has always been done)

**2** Experts (aided/augmented with advanced technologies)

**3** Networks/communities of experts
(e.g. medical specialists, flash teams)

local . regional . global

**New & Non-Traditional Experts**

**4** Experts (with non-traditional qualifications e.g. nano-degrees)

**5** Para-professionals (often aided/augmented with advanced technologies)

**6** Users (augmented with smart self-help systems, and/or with a network/community of other users)

local . regional . global

**Machines & Algorithms**

**7** Fully automated (based on human-designed parameters but no human intervention e.g. automated alerts, IoT)

**8** Machine generated (TASK originates from machines e.g. deep learning)

**9** "Turing technologies" (i.e. performs like an autonomous human, we can't tell difference)

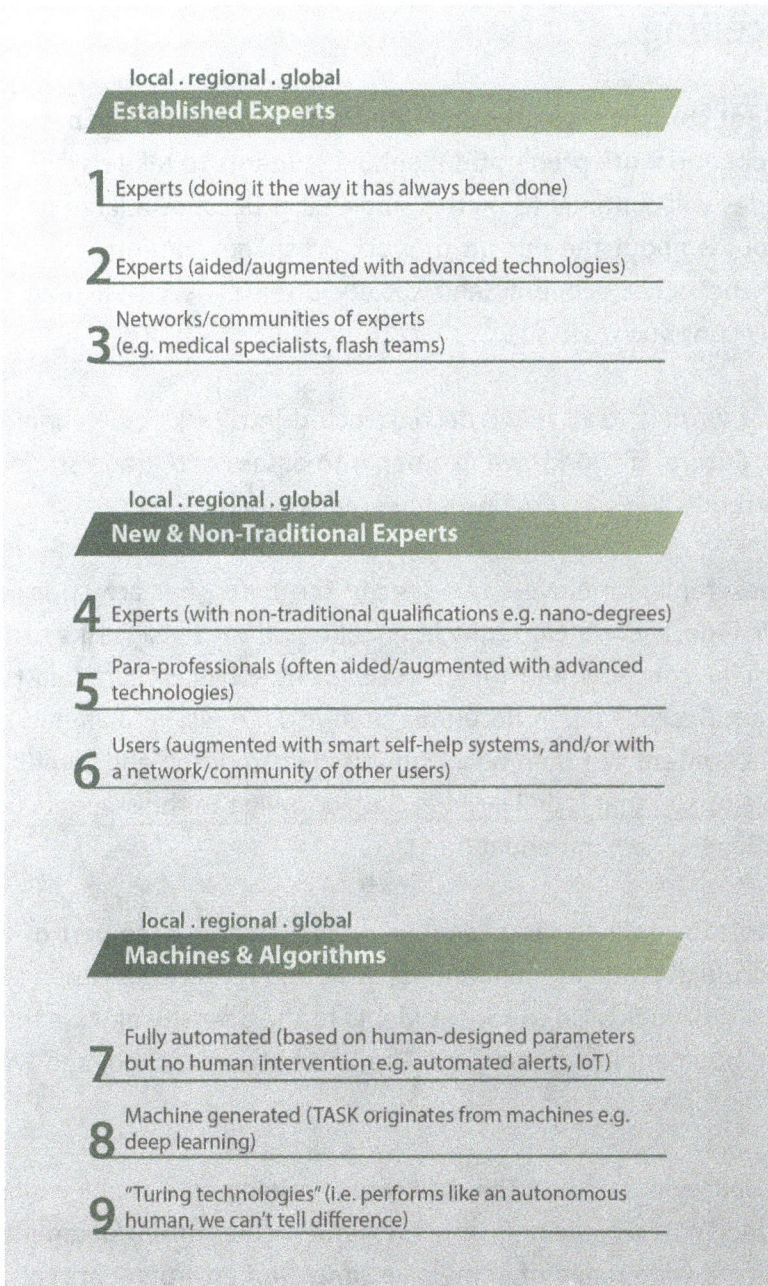

Fig 5.2.5 — Where a Global Team of Mentors for Students and Teachers Could Come From

# Excelling

**8) Spot student strengths and talents by erasing the line between school and work preparation: enable students to take risks**[75][76]
Grades will continue to matter, and that is because a large part of school is about the pursuit of work prospects. Parents, students, and the wider economy and society often expect that (and will likely continue to do so).

But if work is itself being deconstructed into tasks by technology (see Future of Work), we can begin to assess and grade students differently. We can redefine what grades mean.

For example, employers can specify the tasks that are critical to their industry. We can draw connections from these tasks to the modular concepts and topics in different subjects' curricula (see Lesson Design Map in Recommendation 1). A lesson plan and relevant content can then be automatically composed and combined into a lesson that is delivered in class or online to the students (see EduBang in Recommendation 1).

Do this for enough tasks, and we would cover a large part of the curricula deemed as foundational basic and tertiary education subjects. We would also expose students to the diversity of ways these have been applied in industry. The line between school and work has disappeared.

We can even conduct the classroom training as a training simulation. With advances in graphics and virtual/mixed/augmented reality, a wide range of immersive simulated environments can be created for a wide variety of tasks. These can be used to give practical hands-on training and assessment for students. They will be able to develop strong academic, applied, and hands-on skills. And the training can be tailored with the right combination of academic, applied, and hands-on skills, according to the strengths and needs of the students.

Thus when students complete their assignments or take the exams for these lessons, they will essentially be graded on how well they understand how the academic concepts are used in the real world. At one level, one could argue this is taking vocationalisation of education to an extreme. At another, it helps build respect among students for various professions e.g., caregiving tasks for example could be lessons in biology (see Future of Health). At an even higher level, it is a compelling introduction to the ingenuity of humans in applying the sciences, social sciences, arts and more to create innovations and improvements for the world.

There are additional benefits. This will further enhance career counselling in schools to help students make better future career choices. Teachers will be able to spot where students might have unique talents and strengths, and where they might have weaknesses, and use that to guide the students further, both for school and work. Employers will also be able to match work needs better: the good grades map directly to work requirements. And if two students have the same grades but in different tasks, employers will develop a more sophisticated appreciation of the two.

All these could also mean that when it comes to work prospects and preparations, schools can become more focused, efficient, and

**Education and Careers: The Good, the Bad and the Ugly**

In two focus groups with current and recently graduated students from MIT and the Singapore University of Technology and Design, they had some interesting ideas for career advice, preparation and exposure.

They shared that career-wise, students could be categorised into:
- those who know what they want
- those who think they know what they want
- those who don't know what they want but think school can help
- those who don't know what is going on

To cater to all four categories, the focus groups pointed out that they did not just want to see and experience the "good stuff". They also wanted to know about the "bad stuff" because that is a "big decision point" too.

This is an area where taking part in career and training simulations could help. The moderators and participants concluded semi-jokingly that such simulations could be structured and titled to expose them to the good, bad, and ugly of different jobs e.g.
- "Being an Engineer: I Spring Out of Bed Every Morning!"
- "Being an Engineer: I Should Have Stayed Home!"

Fig 5.2.6 — Erasing the Line Between Education and Work - Making Better Choices

streamlined. Valuable time might even be freed up and used to encourage students to invest in their strengths and peaks (see Recommendation 7). Or to pursue curiosity. Assured that the grades map well to the real world, thus improving their odds of securing work when they leave school, they might be prepared to take more risks and try new and different things in school and life.

## 9) Be agnostic about measures of success: grading what you are good at

Armed with a better understanding of their strengths, and assured about their chances of employment, students will have a better sense of what they are good in. They can then strive to hit their peak in those areas. It is likely that different students will be good at different things; thus how we nurture and define success will have to change.

Schools will have to celebrate successes of different stripes. They will have to define success on their students' terms. They will have to become agnostic about different measures of success. A CEO we interviewed suggested "following the child's calling" instead of "following the curriculum" as the future we should aspire to. This is after all what parents hope for their children. That was what one parent asked that we explore as she left one of our workshops. The voices advocating for schools to meet these aspirations will only grow louder. As it is, there is already a trend of increasing acceptance of a greater diversity of study and career choices, including what would have been considered non-mainstream even just ten years ago.

## Conclusion: Building an Education Ecosystem — creating city-scale opportunities

The above recommendations essentially form an education eco-system. Technologies, infrastructure, interactions, networks, and communities will all have to be built. They create both social and economic opportunities. Communities help schools, and schools help communities. Each party in the ecosystem collaborates with each other. Each party strengthens each other. Each party helps each other innovate and become better.

And the entire ecosystem is both local and global. When we teach the world, and the world helps us, the ecosystem accesses and as-sesses the world's needs, resources, and collective intelligence. It could attract the best ideas, innovations, and technologies world-wide.[82][83] The ideas, innovations, and technologies developed within the ecosystem could also travel back out to the world.

The ecosystem serves our education needs first and foremost. It has a public and social purpose. Do that well and - intriguingly - it could also create an innovation and economic ecosystem. The ecosystem can strengthen the social capital and economic future of the city.

## 5.3 Education in 2040

## The (Fun)ctional School
*School had become fun because fun had become functional*

Even the fun advocates had to agree with their strongest critics. School had become fun, not because everyone thought it was good to have fun, but because fun was now considered an essential part of the functional training of students.

This was borne out of necessity. The competition amongst cities and companies had ratcheted upwards. Staying ahead meant constantly innovating. It was not just business and enterprise innovation. It was also social, political, and artistic innovation. Whether you measured it by GDP or GNP, by liveability or lovability, or by happiness or healthiness, there was no sector — public, private, or people — in the city that did not seek or need innovation.

These realities meant innovation had become part of the functional training of school. The focus of school shifted as a result. Getting good grades was great. But what was even better was to show you had innovated.

It quickly dawned on everyone that this meant one thing. You cannot just mandate innovation. You had to start with individual interests, and choose to persevere and excel in it. All these had to be nurtured. Regardless of whether it was for career prospects specifically or for life in general, each and every student had to be given the chance to become very good or even the best in what they were strong in.

The best way to do this? Make school as interesting and fun as possible. Fun as the starting point for life-long innovation had become functional.

Not that the fun advocates minded. They had some concerns that fun had been vocationalised; but for now, all that mattered to them was that fun had a voice in education. Not that the students minded. Why would they, when classes were so fun, so exploratory, and so immersive? Not that companies minded. Well, as long as fun meant more innovation and there was a return on investment on fun.

> Schools were now spotters of individual strengths. Poor teachers though — they were constantly chasing changes. Poor parents too — they found they had turned into "drone parents".

Schools were now spotters of individual strengths and talents. Marshals of mentoring teams for each student. Curators of the best education and entertainment content and technologies. Bridges across socioeconomic and digital divides. And anchors of city-scale and global-scale communities of learning.

Poor teachers though. They were relentlessly being upgraded and constantly chasing changes: their roles in and out of the classroom, how students learned and were taught, and the technologies that enabled and made all this possible. It was breathtaking but breathlessly so.

Poor parents too. This was all so different from the time they were in school. It was tough to keep up. They were nonetheless pleased to see how curious and enthusiastic their children were about school and the world. They were also pleased that their children had a better understanding of the world and not just the syllabus. But they still could not help being anxious — they just wanted the very best for their children. Once children of "helicopter parents", they now found they had turned into "drone parents".

Poor schools then. Parents constantly needled the schools about how their children were doing (much to the schools' consternation). The upside of the parents' anxiety was that many of them offered to help. Parents and schools had learnt to work with each other's strengths.

It was a very (fun)ctional partnership.

# Artefact from the Future: Lesson Design Map

**Juan,** late 40s
Parent
volunteer

## Year 2020: Lesson Design Map[77]

*Deconstructing syllabi to nurture synthesising minds — integrating connections across school subjects, global citizenship, values, and everyday life.*

Juan had that funny feeling. The one where all seemed to be working well, but you feel something is amiss.

That was how Juan felt about his children's education.

His children were doing alright in school (he wished they did better, but which parent did not?). But he was worried that they were not seeing connections between their classes. They could answer the exam questions for each subject, but struggled to draw links between different classes. This was apparent whenever he asked his children about what they were doing in project work. They always needed handholding.

Juan knew the value of seeing connections. He saw it all the time at work. The most innovative ideas often came from insights at the boundaries of different disciplines. The new exciting field of smart fabrics, for example, spanned across the disciplinary boundaries of fashion design, materials, manufacturing, and engineering. Boundary spanning was important if companies wanted to beat their competition and succeed. He often joked that that there were no new ideas, only new combinations.

Juan had read Harvard professor and development psychologist Howard Gardner's *Five Minds of the Future*. In that book, Professor Gardner wrote that to thrive in the future, five intellectual approaches were needed:

1) The disciplined mind was about "mastering at least one way of thinking"
2) The creating mind "breaks new ground"
3) The respectful mind "welcomes differences"
4) The ethical mind "serve[s] purposes beyond self-interest"
5) The synthesising mind "takes information from disparate sources ... and puts it together in ways that make sense"

He felt school was doing alright for the first four minds. But as for the fifth, he was not so sure. He agreed with Professor Gardner that "the capacity to synthesise becomes ever more crucial as information continues to mount at dizzying rates". He felt that developing a flexible mind comfortable with multiple disciplines was something that should start in school.

His daughter Jaz also suggested drawing connections between schoolwork and everyday life. Seeing relevance in what she learnt in school was arguably her number one wish. She longed to be able to explore her own pet interests, such as fashion, as part of her coursework.

Juan decided to do something about it. To the consternation of Jaz's teachers and principals, he became that proverbial pesky parent. He incessantly asked the school "how can we nurture an inter-disciplinary mind in the classroom?"

He bugged them to no end about it. To his credit, he volunteered to help. To the consternation of his friends, he "volunteered" them as well. To the dismay of Jaz, he "volunteered" her help to the school too, over and beyond her academic and co-curricular activities (thanks Dad...).

After several rounds of discussion, they came up with a Lesson Design Map (see Fig 5.3.1). The teachers and friends who were "volunteered" (albeit with some kicking and screaming) had to agree that they liked the Lesson Design Map.

Teaching multi-disciplinary classes had always been a challenge. The Lesson Design Map was not perfect, but the teachers felt the handy guide was a good start. At the very least, it pointed to specific existing curriculum concepts and topics across current subjects.

This ensured rigour. It also provided helpful links, saving them a tremendous amount of time. And lastly, there was no need to learn yet another new piece of technology, as no additional technology was needed.

The teachers also liked the idea they might be nurturing future climate change innovators. They knew that what happens in the classroom is seminal in seeding ideas, shaping values, and setting aspirations. That was why they got into teaching in the first place. The earlier students understand the multi-faceted and connected challenges of climate change, the greater their capacity to tackle the challenges of climate change 5, 10, 20 years later.

Juan's "volunteered" friends were happy that they might have contributed to the future of education and even the future of the world.

Juan was of course pleased and grateful. He had worked very hard on this. But Juan was also unsure if this could be sustained. Everyone was so busy these days. He had felt bad at times about having gotten the teachers, his friends, and Jaz involved in this "extra-curricular" pet pursuit of his.

If only Juan knew what the Lesson Design Map would unleash.

# LESSON DESIGN MAP
## FOR A SYNTHESIZING MIND
Schematics of Teacher's Pull-Out Map

**Question & Objective**
The big questions and learning objectives addressed by the school subjects.

**School Subjects**
Different subjects are color coded. Existing syllabi/curricula topics and concepts; anchoring it on what already exists ensures rigor, familiarity (schools have been using them), and saves work and time for teachers and students (no need to reinvent the wheel).

Geography

Chemistry

**01**

**What is climate change?**

. O B J E C T I V E .
The knowledge drawn from Geography and Chemistry can explain the greenhouse gas effect. This provides background to the economics of climate change. It is also useful in the designing and evaluation of policies.

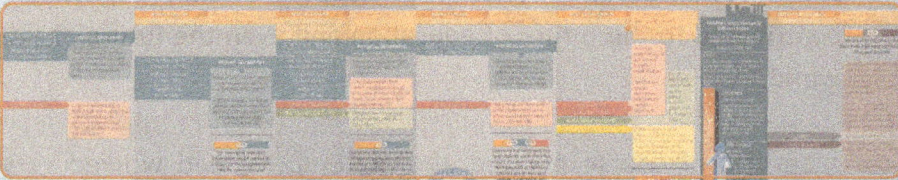

**Topics & Explanations**
Grouping different subjects' topics and concepts within each big question helps students see connections.

**Case Study**
Case study links classroom to world outside, helping students see additional connections.

Fig 5.3.1 — Lesson Design Map - *Enlarged versions on next two pages; complete map can be downloaded at* https://livingdigital2040.files.wordpress.com/2016/09/lesson-design-map-web.pdf

# LESSON DESIGN MAP
## FOR A SYNTHESIZING MIND

Subjects :

| Economics | Geography | Chemistry | Physics | Biology | Ethics | Literature & Arts |

---

## 01

**What is Climate Change?**

**. O B J E C T I V E .**
The knowledge drawn from Geography and Chemistry can explain the greenhouse gas effect. This provides background to the economics of climate change. It is also useful in the designing and evaluation of policies.

## 02

**Who is responsible?**

**. O B J E C T I V E .**
Students need to be aware of the different perspectives towards this issue. This is strongly encouraged in the new A Level Economics syllabus (2016).

## 03

**Why is it happening?**

**. O B J E C T I V E .**
Students can draw knowledge from other disciplines such as Geography, Chemistry and Physics. This adds more breadth and depth when they explain economics concepts (eg. negative externality).

---

| Topics | Explanation & Links | | Topics | Explanation & Links | | Topics | Explanation & Links |

**MARKET FAILURE** — Economics Topic / Economics Concept

**MARKET FAILURE** — Economics Topic / Economics Concepts

### PHYSICAL GEOGRAPHY
ATMOSPHERIC PROCESSES
HAZARDS & MANAGEMENT
VARIABLE WEATHER
CHANGING CLIMATE

Climate change refers to changes in average weather conditions over an extended period of time due to excessive production of greenhouse gases.

CHEMICAL BONDS — Chemical reactions arising from industrial and agricultural activities contribute to greenhouse gases such as CO2, CH4 (methane) and N2O (nitrous oxide).

### HUMAN GEOGRAPHY
GLOBALISATION OF ECONOMIC ACTIVITY
POPULATION ISSUES & CHALLENGES
URBAN ISSUES & CHALLENGES

Anthropogenic climate change can be discussed in all topics across Human Geography.

Arguments against the anthropogenic climate change often refer to the Milankovitch cycles which argue that global warming is natural.

The concept of imperfect information can be applied to account for the disagreements over the causes of global warming.

### PHYSICAL GEOGRAPHY
ATMOSPHERIC PROCESSES
HAZARDS & MANAGEMENT
THE GLOBALISATION OF ECONOMIC ACTIVITY
POPULATION ISSUES & CHALLENGES
URBAN ISSUES & CHALLENGES
ORGANIC CHEMISTRY
SPECIFIC HEAT CAPACITIES

Enhanced Greenhouse Effect traps more outgoing radiation leading to excessive global temperatures.

Greenhouse effects from carbon monoxide, oxides of nitrogen and unburnt hydrocarbons. Ozone depletion from fluoroalkanes and CFC.

Compare greenhouse gases leaves trapping heat vs other gases.

Geography provides an overview of key categories associated to human activities that contribute to global warming — this can be a helpful framework to organize examples when approaching the concept of negative externality.

Chemistry and Physics provide specific scientific examples behind the economic concepts.

---

## A CASE EXAMPLE: FAST FASHION

**What is Fast Fashion?**

RUN WAY

RETAIL

**What does it have to do with Climate Change?**

FAST PRODUCTION

ACCELERATES environmental pollution

**2nd** LARGEST INDUSTRIAL POLLUTER

We are all responsible. How much do our consumers care in terms to find out how our clothes are made?

Do companies try to provide that information worth their consumers who choose to produce using environmentally friendly processes?

Lastly, do governments regulate these activities?

**WHY IS IT HAPPENING?**

# LESSON DESIGN MAP
## FOR A SYNTHESIZING MIND

**School Subjects :**

| Economics | Geography | Chemistry | Physics | Biology | Ethics | Literature & Arts |

**04** What is the impact to economy & society?

. O B J E C T I V E .
Students can draw from other disciplines such as Geography and Chemistry to explain deadweight loss from negative spillovers and who will be affected.

**05** What can be done?

. O B J E C T I V E .
Students will be introduced to economic instruments and technology solutions that can tackle climate change.

**06** Why should we care?

. O B J E C T I V E .
In conclusion, students encouraged to reflect the purpose of learning Economics in the light global challenges facir whole of humanity suc climate change.

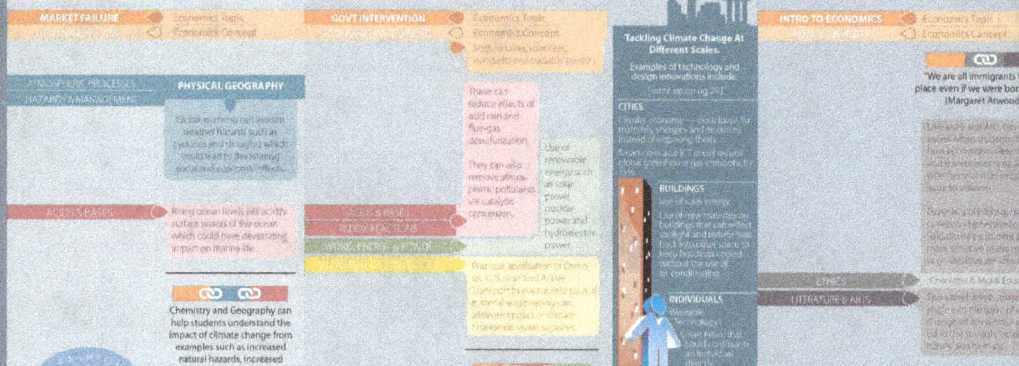

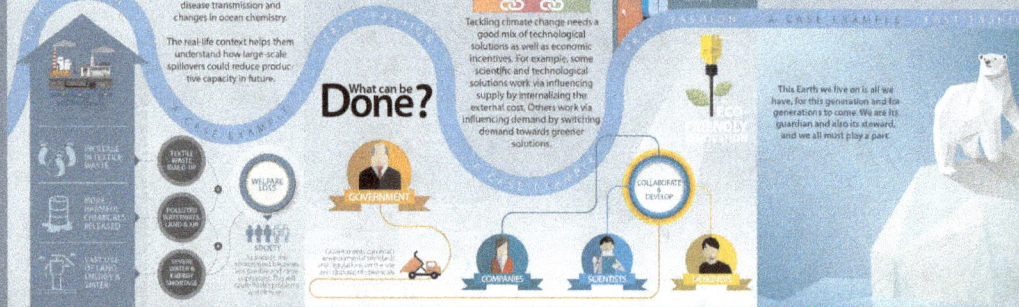

Fig 5.3.1 — Lesson Design Map

*Enlarged version of the complete map can be downloaded at https://livingdigital2040.files.wordpress.com/2016/09 lesson-design-map-web.pdf*

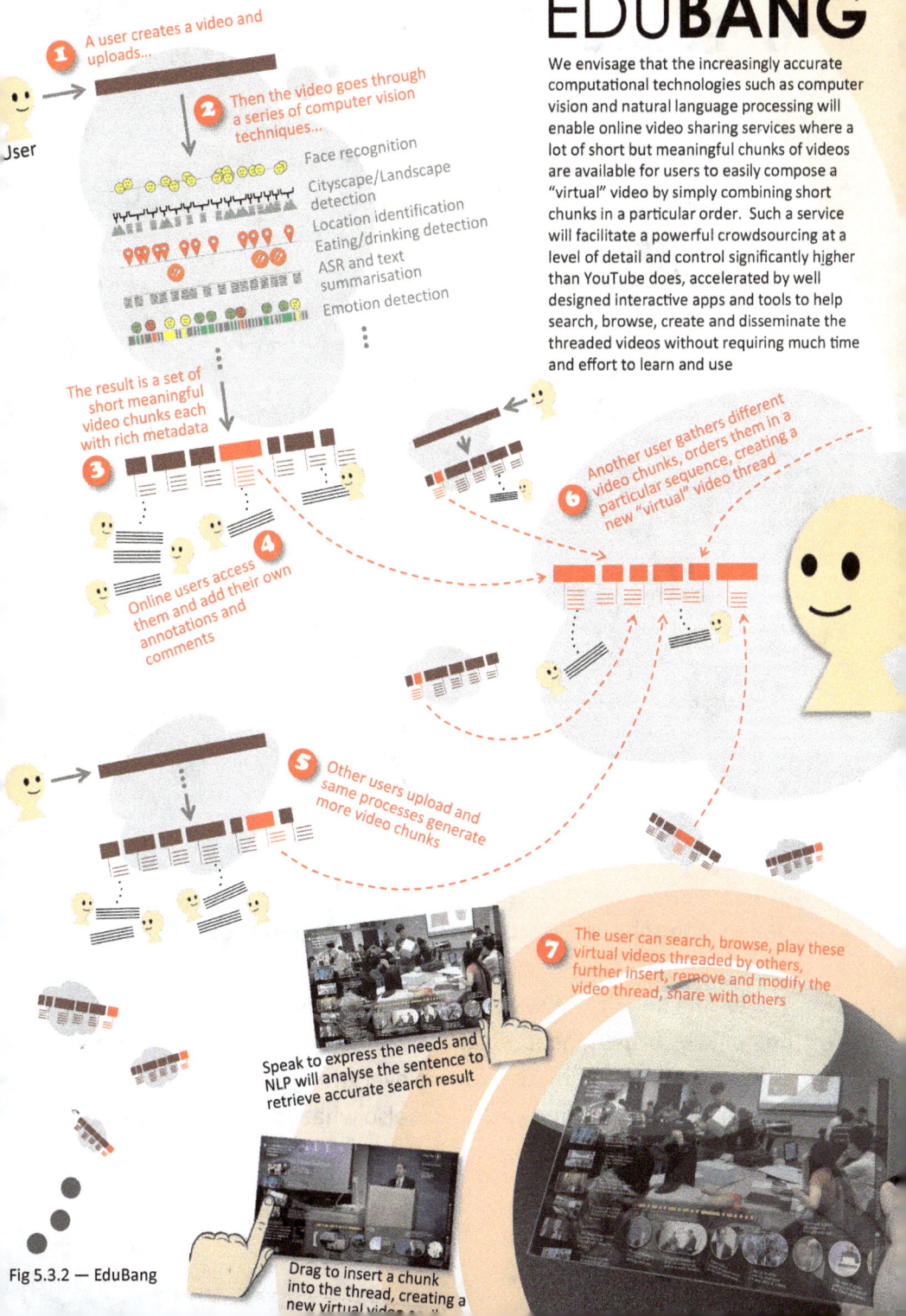

Fig 5.3.2 — EduBang

**Artefact of the future circa 2025**

Poster presented at a conference in Human Computer Interaction and Education

see video at:
https://youtu.be/8QkqNnKIGPA

# EDUBANG

We envisage that the increasingly accurate computational technologies such as computer vision and natural language processing will enable online video sharing services where a lot of short but meaningful chunks of videos are available for users to easily compose a "virtual" video by simply combining short chunks in a particular order. Such a service will facilitate a powerful crowdsourcing at a level of detail and control significantly higher than YouTube does, accelerated by well designed interactive apps and tools to help search, browse, create and disseminate the threaded videos without requiring much time and effort to learn and use

1. A user creates a video and uploads...

User

2. Then the video goes through a series of computer vision techniques...

Face recognition
Cityscape/Landscape detection
Location identification
Eating/drinking detection
ASR and text summarisation
Emotion detection

3. The result is a set of short meaningful video chunks each with rich metadata

4. Online users access them and add their own annotations and comments

5. Other users upload and same processes generate more video chunks

6. Another user gathers different video chunks, orders them in a particular sequence, creating a new "virtual" video thread

7. The user can search, browse, play these virtual videos threaded by others, further insert, remove and modify the video thread, share with others

Speak to express the needs and NLP will analyse the sentence to retrieve accurate search result

Drag to insert a chunk into the thread, creating a new virtual video

# 6   Future of Healthcare

Fig 6.1.1 — Future of Healthcare Scenarios

## 6.1 Scenarios[1–101]

Health in the city was starting to look like a Jackson Pollock painting.

The formal healthcare system — hospitals, clinics, and going to the doctor's — firmly anchored the city's health system, but formed only a slice of it. Health was now in everything. It was in you. It was on you. It was all around you. It was also what you put into you. And where you took you.

Digital had made health omnipresent. It sensed, measured, and monitored everything you did and ate, and everywhere you went. At home. In offices and co-working spaces. Buildings and bars. Even in the ABCs (autonomous buses and cars) you commuted in. It then analysed all of this, and told you what it meant for your health, given your genes and the environments you were in. In real time, live-streamed, 24/7/365, if you so desired.

Digital health was both omnipresent and omnipotent. There were hence frequent calls for well-integrated technologies. But these also posed privacy and security risks. At the same time, all the advanced technologies mattered little if citizens were not motivated to manage their own health.

Digital health felt omnipotent. Knowing what it all meant was just the beginning. Cities, companies, and citizens could do so much more now than jiggle fitness trackers. From omics technologies, to nanobots in the body, to medical body area networks, to robo-medical professionals (equipped with oxymoronic sounding "artificial intelligence bedside manners" no less), to urban scale genome-wide association studies, to cross-continent telehealth, there were technologies of all stripes. They detected, diagnosed, and defeated diseases. Where they could, they also prevented diseases. Digital health added years to life, and life to years. It really seemed like there was nothing digital could not do.

There were hence frequent calls for technologies to be well integrated. The more integrated they were, the more complete the health picture. And the more cities, companies, and citizens could do for and with each other.

Ominously, it also meant the more that cities, companies, and citizens could <u>do</u> <u>to</u> each other.

The more integrated the systems were, the more citizen privacy was at risk. Citizens worried about what companies and cities would do to their personal health data and information. The more integrated the technologies and systems were, the more vulnerable the technologies were to system-wide cascades of cyber failure, fraud, and felony.

And then there was the matter of the citizens' motivation. You can lead a horse to water, but you cannot make it drink. All the personalised technologies, information, and data mattered little if citizens were not motivated to manage their own health. Look at diet and lifestyle for example. Citizens have known for a long time they should eat less and move more. Many chose – consciously, subconsciously, and unconsciously – instead to eat more and move less. It was not enough to know. They had to also want to and actually do the right thing.

The future of health was thus determined by how motivated and able people were to manage their own health, and how integrated or partitioned the technologies were. Where cities, companies, and citizens were on these dimensions determined the future of healthcare.

Fig 6.1.2 — Future of Healthcare Scenario - Nag

# Nag

The "Nag" cities believed in integration and that citizens needed to be helped actively.

There was no other way because the mind was willing but the body was weak – because of their busy schedules, citizens simply found it difficult to manage their own health and hence became less motivated to do so. So the cities decided to invest in and integrate all the technologies needed. They could then use the integrated system to provide personalised healthcare, and to send out constant just-in-time nudging advice and actions to individual citizens.

Only that citizens began to feel that they were being nagged all the time. There was a fine line between nudging and nagging.

This was because the cities' integrated systems took care of everything. All the information about your health — genomic profiles, clinical records, research informatics, diet, home and work environments, commutes, cross-border travels, etc. — were automatically fed into this mega digital brain. It had a ready analysis and recommendation for everything you did.

Companies were pleased with this system. It took a substantial chunk of employee health-related costs off their books. All they had to do was to feed relevant information into the mega digital brain. Very plug-and-play.

Citizens did not entirely mind that this was quite intrusive into their privacy. Despite initial reservations, they soon considered it a necessary trade-off, even evil, against the benefits of receiving personalised healthcare. The city was taking care of them; surely there was no need to complain like a champion grumbler?

Well, there was. Privacy was one thing. Nagging, another. Citizens often felt that this mega digital brain "doth protest too much". Over time, more and more citizens developed an immunity to the nagging, choosing to ignore reminders such as "that extra piece of fried lab-grown chicken will tip your weekly fats target".

Sometimes, people just want to have an extra piece of fried chicken. Guilt-free. And nag-free.

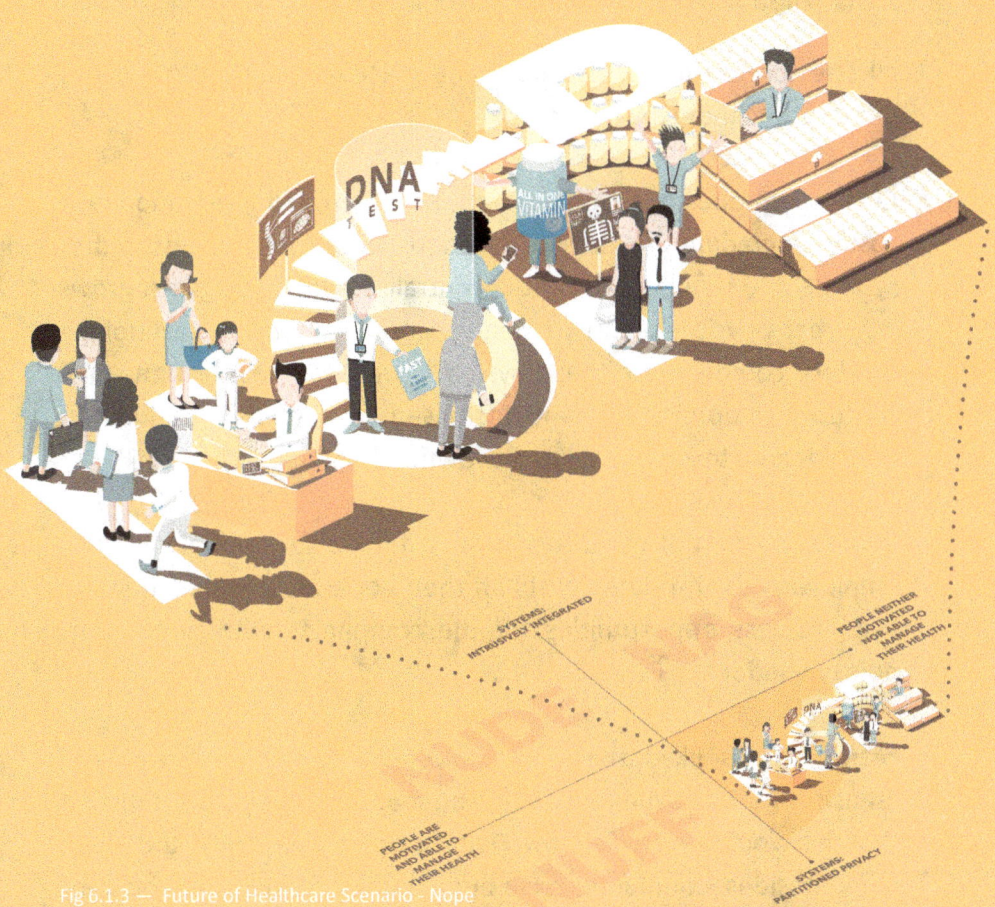

Fig 6.1.3 — Future of Healthcare Scenario - Nope

SYSTEMS: INTRUSIVELY INTEGRATED

PEOPLE NEITHER MOTIVATED NOR ABLE TO MANAGE THEIR HEALTH

NODE

NAG

NUFF

PEOPLE ARE MOTIVATED AND ABLE TO MANAGE THEIR HEALTH

SYSTEMS: PARTITIONED PRIVACY

# Nope

The "Nope" citizens said nope to having to do more. Because of their hectic schedules, citizens simply found it difficult to manage their own health, and were not motivated to do more.

They also said nope to integration. After the Great Data Debacle, they had been wary of putting any personal data and information in any one system or organisation. Citizens prized privacy above all else, preferring to distribute and partition information across different technological systems.

Cities said nope to integration too. For them, full integration was costly; upgrading the current partitions would suffice. Distributed systems also offered a default decentralised defence against growing cyber-physical security threats. Moreover, making technology vendors compete for these systems promoted innovation, prevented monopolisation, and increased market competition. All these helped to keep total costs down.

Companies did not mind this at all. It took a substantial chunk of employee health-related costs off their books. It also meant lucrative business opportunities for the companies who were technology vendors.

Many citizens liked the status quo. The current nudges were "satisficing".[76] Maybe there could be more, but they were satisfied that this would suffice for now. Doing more felt like a little too much work. Besides, companies have been touting for decades the benefits that integrated systems would bring, but where's the beef? Citizens felt they were doing just fine at the moment, making steady incremental improvements.

Other citizens were not so sure. The cost of technology was falling all the time and the case for including new data and information was growing in tandem with the latest research. Was it time to seek a second opinion about the future?

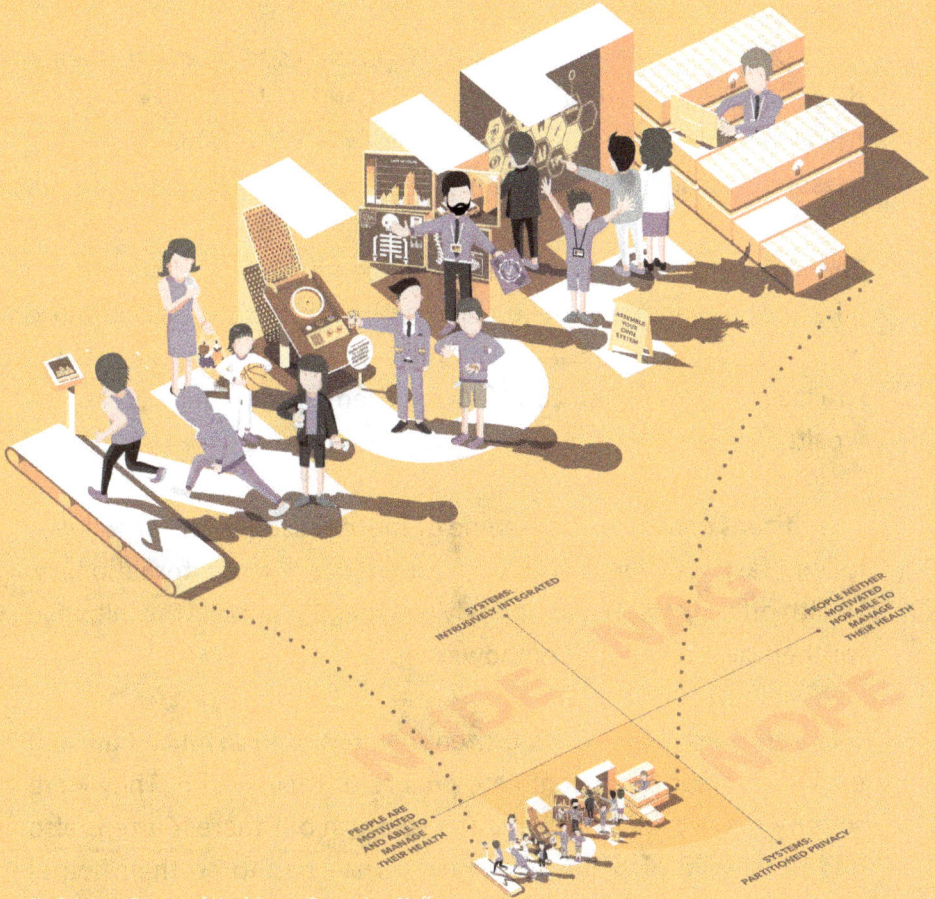

# Nuff

The "Nuff" group of cities believed in providing just enough. Leave it to citizens and companies to lead the way. Citizens were thus forced by circumstances to take responsibility for themselves, using a wide range of technologies that were often not integrated with each other.

Cities had good reasons for doing this. Technologies had advanced so rapidly, that what citizens could access often outpaced what cities could provide. Be it cheap DIY genomics tests, multi-disease diagnostic mobile tricoders, SuperSuits, robo-doctors, or international tele-medicine services, citizens could assemble their own personalised healthcare system.

The logic was this: where citizens go, companies will follow. After all, citizens tended to spend more on health as they earned more and also as they aged. And companies were eager to get access to citizens' data. They could turn citizens' health into corporate wealth.[77–91]

The citizens concurred with some of these reasons. The market was indeed faster. Expertise could be found locally and internationally. Health offerings were increasingly cheaper. And taking the initiative for their own health felt empowering.

But there were side effects. Citizens had to spend so much time and energy to assemble a complete picture of their health. They were so busy these days, and there was so much out there. Citizens also had to beware of scams and cheats. They had to do their homework to separate the wheat from the chaff.

There was a creeping social cost too. Individual technologies were becoming more affordable, but the costs escalated once you tried putting them together comprehensively. There were already signs of health divides in the city. They risked turning malignant.

Citizens also began to think, if I have to do so much for myself, why should I pay so much taxes to fund the public healthcare system? And if I am healthy, why am I not paying a much lower insurance premium? Let those who are not so healthy and who do not take care of themselves pay more. I want to pay just enough. Nothing more, nothing less, 'nuff said.

Was DIY, *caveat emptor*, and fading shared social responsibility really the best way to ensure good health in a city?

Fig 6.1.5 — Future of Healthcare Scenario - Nude

# Nude

Citizens in "Nude" felt absolutely naked.

The integrated view of their health with interventions tailored to their motivation levels empowered them to take charge. But everything about their health could also be shared with cities, companies, and communities. There was nowhere to hide, and they felt quite exposed.

The cities had decided that integrating systems and technologies was critical to public health. They also knew that only they were capable of providing the right combination of cybersecurity, policies, and legislation to make this happen. And only they could corral the companies to work with them for the public good.

As a result, citizens found it easy to take responsibility for their own health. It did not stop there. With health, information and data democratised and decentralised, citizens created community initiatives to help each other. Want to know if this diet is suitable for your genomic profile? Here's a group in your city that's just like you. Want to know what it's like to be diagnosed with a rare ailment? Here's someone overseas who has been there, done that. Everyone had become a helpful health nano-expert to someone else in the city and the world.

It all seemed very ideal. But one big wrinkle in all this was cost. Building and maintaining such integrated, complex, and secure systems and technologies was very expensive. Cities, companies, and citizens had worked out a model of sharing the costs. But the nature of healthcare meant costs were likely to continue rising. Might the business model break down at some point?

Another big wrinkle was privacy. Big Data meant risks of Big Theft. And it was one thing to have Big Brother watching you; but now you had Big Father, Mother, Sister, Spouse, Children, Cousin, Nosey Neighbour and Fussy Friend watching too. All this spelt potential Big Trouble.

At the same time, democratisation shifted power from the existing experts and established order in healthcare. Now that every citizen was a nano-expert, they were constantly challenging the health-care system and its professionals. Was this a case of "a little knowledge — augmented with technology — is a dangerous thing"? Or were they simply symptoms of an evolving robust and healthy relationship between people and professionals?

The naked truth was out there.

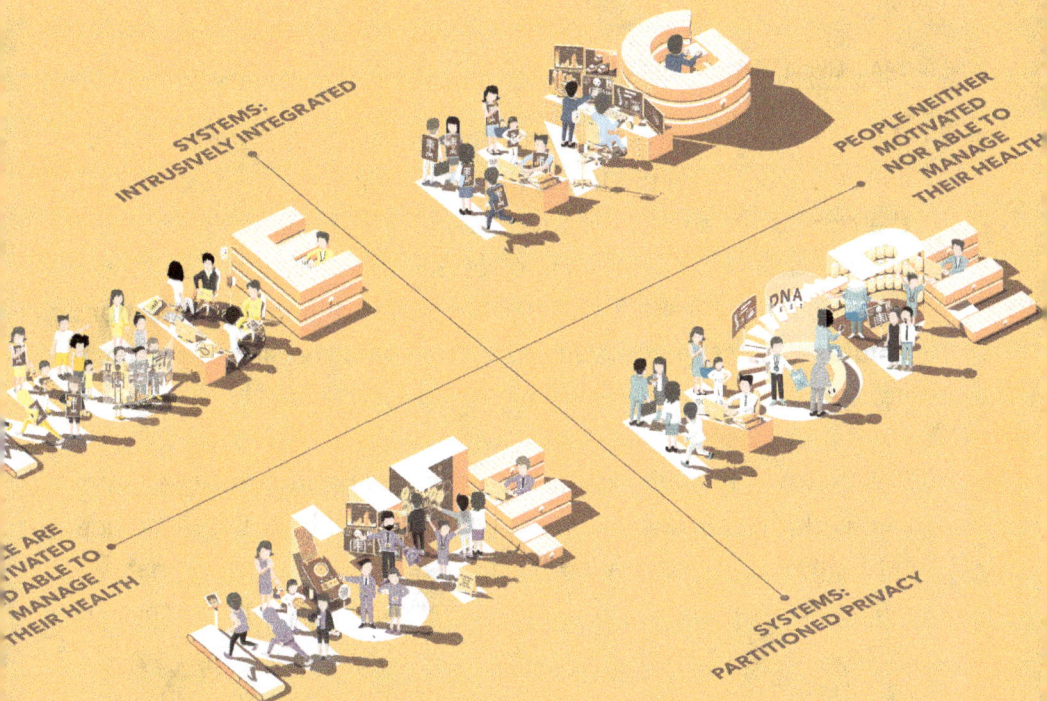

SYSTEMS: INTRUSIVELY INTEGRATED

PEOPLE NEITHER MOTIVATED NOR ABLE TO MANAGE THEIR HEALTH

DNA

...E ARE ...IVATED ...D ABLE TO MANAGE THEIR HEALTH

SYSTEMS: PARTITIONED PRIVACY

Fig 6.2.1 — Responding to Future of Healthcare Scenarios

## 6.2 Recommendations

The future of healthcare will be anchored by the healthcare system, but it will be larger than the healthcare system. As a hospital head described, today's healthcare system will first evolve into a community-based **healthcare** ecosystem. It would then evolve into a community-based **health** ecosystem.

Our interviewees shared that the ideal case would be one where all technologies and systems are integrated. They acknowledged however this would be a long term endeavour. One doctor said it might only be in 2040, the target year of *Living Digital 2040,* that we will see this level of integration.

We will make progress towards this in spurts and starts. But we will zigzag between the four different scenarios at various points in the coming decades. Whichever scenario it is, one thing was clear across our interviews: technologies will take over more and more tasks, but we will still have a large number of human interactions in future. These might even grow with the increasing use of technology.

But that was not the most interesting finding that emerged from our work. The most intriguing insight is this: with digital, we do better individually when we do better together. If we can do that, we will be ready for any of the four scenarios, no matter how they zig or zag.

Our recommendations thus revolve around this **unifying idea: we take better care of ourselves when we take better care of each other.** And we can do so in the following ways:

- empower people
- energise community and society
- elevate how people and professionals interact

| Approach | Recommendations |
|---|---|
| **Empower People** | 1) Identify motivation and activation levels: go beyond health screening and literacy.<br><br>2) Empower people to act on their data and information: make it easy to access and integrate – from dashboards and SuperSuits to policies and regulations. |
| **Energise Community and Society** | 3) Make it social: we can take better care of ourselves when we take better care of others.<br><br>4) Mobilise resources socially: make it easy to scale community resources.<br><br>5) Become a city that cares and gives care: nurturing values for a community-based ecosystem. |
| **Elevate Interactions** | 6) Elevate the professional-patient relationship: make it an adult-adult partnership.<br><br>7) Elevate the professional-professional relationship: strengthen – not lessen – how they work and learn together.<br><br>8) Elevate the people/peer/professional-machine relationship: integrate interactions. |

Fig 6.2.2 — Summary of Recommendations for Future of Healthcare — Empower People, Energise Community and Society, and Elevate Interactions

# Empower People

### 1) Identify motivation and activation levels: go beyond health screening and literacy

If people are motivated to take care of their own health, they are more likely to stay healthy. An excellent description of this is "patient activation" which "describes the knowledge, skills and confidence a person has in managing their own health and health care."[102][103]

We can tailor help according to people's motivation levels. Being motivated or "activated" is both different and more than being literate in health matters. Literacy equips one with skills, but skills alone do not guarantee motivation.

This is not just for patients who might have ailments. It is also for anyone who is in the pink of health. According to Professors Judith Hibbard and Helen Gilburt who wrote a report for the UK's The King's Fund, "patient activation is the best predictor of healthy behaviour over a wider range of outcomes".[104]

The key word is "activation".[105] It is both different from and more than being literate in health matters. Literacy equips you with the skills, but skills alone do not guarantee motivation.

Our earlier example in the scenarios on diet and lifestyle shows this clearly. Citizens have known for a long time they should eat less and move more. Many choose — consciously, subconsciously, and unconsciously — instead to eat more and move less. It is more than knowing. Everyone should feel responsible for one's health, and be motivated to and actually do something about it.

Once we know people's motivation and activation levels, we can tailor our help. Different people would for example need different behavioural "nudges":[106] some might need to raise their motivational levels as a start, while others might simply need to improve their health literacy. Some might prefer help with exercise, while others might prefer help with food and diet.

In future then, health screening will no longer be sufficient. People will also have to screen themselves for their activation and motivation levels.

**2) Empower people to act on their data and information: make it easy to access and integrate - from dashboards and SuperSuits to policies and regulations**
Once people are sufficiently motivated, they will need to have the right data, information and advice at the right time.[107–109]

It is currently very difficult for anyone to access all of this, much less to act on them. Electronic health records are an excellent start. They however only cover some of the data about a person's health. They are also often not directly accessible to the individual. The Quantified Self movement takes a step in the right direction. But its dependence on multiple wearables will impede adoption: an entrepreneur in healthcare analytics doubted wearables would be widely adopted, as there is no easy way to sync and integrate across them.

It becomes even more challenging when the amount of data and the number of wearables and sensors rise. Besides data about ourselves, additional data and information could include:

- environment (e.g., weather, air quality, and temperature data collected from sensors/IoT)
- travel (e.g., overseas trips might expose us to new diseases, a recent example being Zika)
- financial (e.g., debt levels and how we spend money could reflect and affect stress levels)
- relationships (e.g., who we hang out with influences our lifestyle choices in food and drink)
- behavioural (e.g., sudden and uncharacteristic phone inactivity could signal depression)

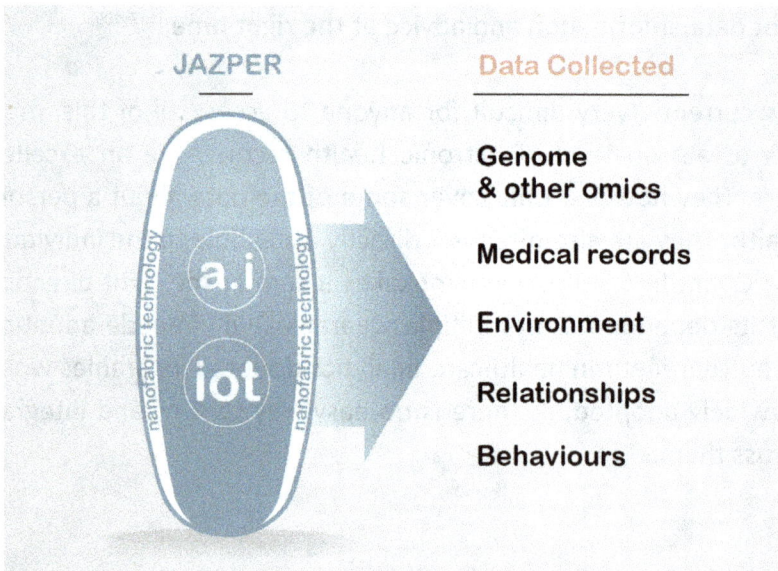

Fig 6.2.3 — Health-related Data in Future (see also Artefact from the Future in section 6.3)

Putting everything together will require several new R&D developments that include:

a) Developing a strong research and clinical basis — for example, Verily Life Sciences (formerly Google Life Sciences), Stanford University, and Duke University have embarked on the Baseline Study, a "comprehensive initiative to understand the molecular markers that are key to health and the changes in those biomarkers that may lead to disease."[110]

b) Developing a Personal Health Intelligence Dashboard — this would not only integrate any new sources of data and information easily, it would also provide sound insights that could be acted on, without overwhelming the user. Good UX (User Experience) will be critical.

c) Developing SuperSuits - one way to overcome the limitations of today's wearables might be to develop the ultimate wearable: the SuperSuit (see section 6.3 — Artefact from the Future). Electro-spun from nanofibres,  and integrated as layers or textile geometry into the clothes we wear, it could incorporate many of the sensors needed.

Fig 6.2.4 — Artefact: SuperCare/SuperSuits/Empathy Suits (expanded picture and details in section 6.3 — Artefact from the Future)

These R&D developments have all been given a recent boost. The Baseline Study mentioned above for example was only started in 2014. The underlying technologies for SuperSuits were given a fillip in 2016 with the US$317 million Advanced Functional Fibers of America consortium involving close to 100 companies and incubators, and 32 universities.[111][112] In fact, two of our interviewees/participants in this field told us many of the underlying technologies are already available, and the next step is to integrate them and make them commercially feasible.

Parallel discussions and decisions will also have to be made on privacy, policies, regulations, ethics, and education. These will have to keep in mind that the pace of changes in the varieties and volumes of data and information will certainly outstrip the speed at which these discussions and decisions take place. We will need to determine how we can make these discussions and decisions dynamic and responsive to the accelerating advances in digital data and information.

We will also need to work out the business models. The report *The Internet of Things for Health Care: A Comprehensive Survey* states clearly that the "business strategy is not yet robust because it involves a set of elements with new requirements such as new operational processes and policies, new infrastructure systems, distributed target customers, and transformed organisational structures...Therefore, there is an urgent need for a new business model."[113]

# Energise Community and Society

**3) Make it social: we can take better care of ourselves when we take better care of others**

Once we have empowered the individuals so that they can do better for themselves, we can energise the community and society to do better together. We can make health social.

MIT Professor Alex Pentland's team at the MIT Media Lab offered an insight into how this could happen through "social network incentives."[114–116] According to Professor Pentland:

> *"Standard economic incentives miss the mark because they frame people as individual, rational actors rather than as social creatives influenced by social ties... There is another way: by providing incentives aimed at people's social networks rather than economic incentives or information that are aimed at changing the behaviour of individual people."*[114]

His team ran an experiment with a community of young families who interacted with each other to varying degrees. They wanted to increase the level of physical activity of the community's individuals.

A typical approach would have been to give each of them an economic incentive to be more active. For example, if the individual hit a target activity level, he or she would get a cash reward. Or they could have been given a group incentive: if everyone achieves a group target, they receive a group reward that can be shared.

But that is not what Professor Pentland's team did. They used a social network incentive instead. Each individual in the community was designated a "behaviour change target". He or she would then be assigned two buddies within the community. If the "behaviour change target" individual hit a target activity level, measured through their smart phones, the two buddies would receive the reward.

Because the team involved almost everyone in this community, each individual was not just a "behaviour change target". He or she would also be a buddy to another "behaviour change target". In other words, each individual in the community played two roles: he could not only earn a cash reward as a buddy, he or she could also help a buddy earn a cash reward.

The social network incentive was four to eight times better than an individual incentive at raising activity levels. Even when the incentive was stopped, the individuals kept up their higher activity levels.

The result? The social network incentive was four times better than an individual incentive at raising activity levels. This increased to eight times between "behaviour change target" individuals and buddies who had more interactions with each other. They also found that even when the incentives were stopped, the individuals kept up their higher activity levels. And contrary to expectations, such social network incentives cost less, even when the "social cost of peer pressure" was accounted for.

We should explore using such social network incentives[117–119] to encourage communities and cities to develop health habits. We might learn that in matters of health, we can take better care of ourselves when we take better care of others.

## 4) Mobilise socially: make it easy to scale community resources

As the healthcare system evolves into a community-based eco-system, more and more of the community resources will need to be mobilised. Both doctors and nurses told us that once patients are out of the hospital, multi-agents have to be involved — social, grassroots, IT providers, social organisations, community, etc.

**The critical question for a community-based health eco-system: can we sufficiently scale up, rally citizens, and mobilise more resources?**

There are existing examples of these. In Singapore for example, the Tsao Foundation has an integrated care system where "civic organisations, health and social service providers, research analysts, policymakers and other stakeholders"[120] collaborate to take care of the elderly in a particular residential district. Another example is the Stanford Peer Health Educators program. Older students volunteer and are trained to "actively promote and support student health, wellness and safety" amongst their peers in the residence halls.[121]

The critical question is this: to be community-based on a large scale, can we sufficiently scale up, rally citizens, and mobilise more resources?

We should continue to promote greater community participation through volunteerism, professional training, community support groups, corporate social responsibility initiatives and charities. We will have to explore new integrated collaboration platforms that can support these — these do not currently exist. We will also have to experiment with collaboration processes e.g., assembling teams on-the-fly to tackle short term challenges, and dedicated teams who can commit to long term initiatives.

We should also experiment with other social network incentives. Another of Professor Pentland's teams tested their use with social media to mobilise resources under time pressure. They found that social network incentives were more effective than altruistic appeals for help: a larger network of help was formed, and more people in the network passed on the requests for help.[117–119]

Mobilising the community will require a multi-pronged strategy. The more tools we have at our disposal, the more likely we will succeed.

### 5) Become a city that cares and gives care: nurturing values for a community-based ecosystem

When we mobilise more in the community, we will increasingly require more and more of them to be competent caregivers. They are doing it as a role (e.g., at home), and not doing it as a profession.

This has two dimensions. The first dimension is social e.g., keeping patients

Can we have a strong community-based health eco-system if the community does not care about caregiving? Just as we encourage citizens to learn basic CPR to respond to emergencies, cities might have to equip their citizens to be able to give basic caregiving. This will also nurture a stronger appreciation of the caregiving professions.

company. That is the focus of many current initiatives. The second dimension is clinical e.g., administering tests and medications to patients. Our interviewees stressed that this dimension is becoming increasingly critical as we  move to a community-based system.

In the same way that we encourage citizens to learn basic CPR so that anyone can respond to an emergency anywhere, anytime, cities might have to equip each of their citizens to be able to give basic caregiving as a role.

We can take advantage of how work is now being deconstructed into tasks (see Future of Work), and the opportunities this offers in schools as a result (see Future of Education). We could, for example, consider taking tasks from caregiving, break them down, and match them to modular syllabus and curriculum topics. These could then be automatically augmented with interesting educational content (see Future of Education — Lesson Design Map and EduBang). They could then be taught in school, potentially with simulation technologies (see Future of Work and Future of Education).

By doing this, we equip every citizen in the city with basic caregiving skills, ensuring that the community always has a wide pool to draw on. Someone is always around to render basic clinical care.

There is a wider benefit. If we are moving to a community-based ecosystem, then the caregiving professions and the act of caregiving in general must be valued. Otherwise how can one have a strong community-based ecosystem if the community does not care about caregiving?

Training every citizen helps them better appreciate caregiving and the caregiving professions. We are not just equipping citizens with skills. We are also nurturing shared values of what the city cares about.

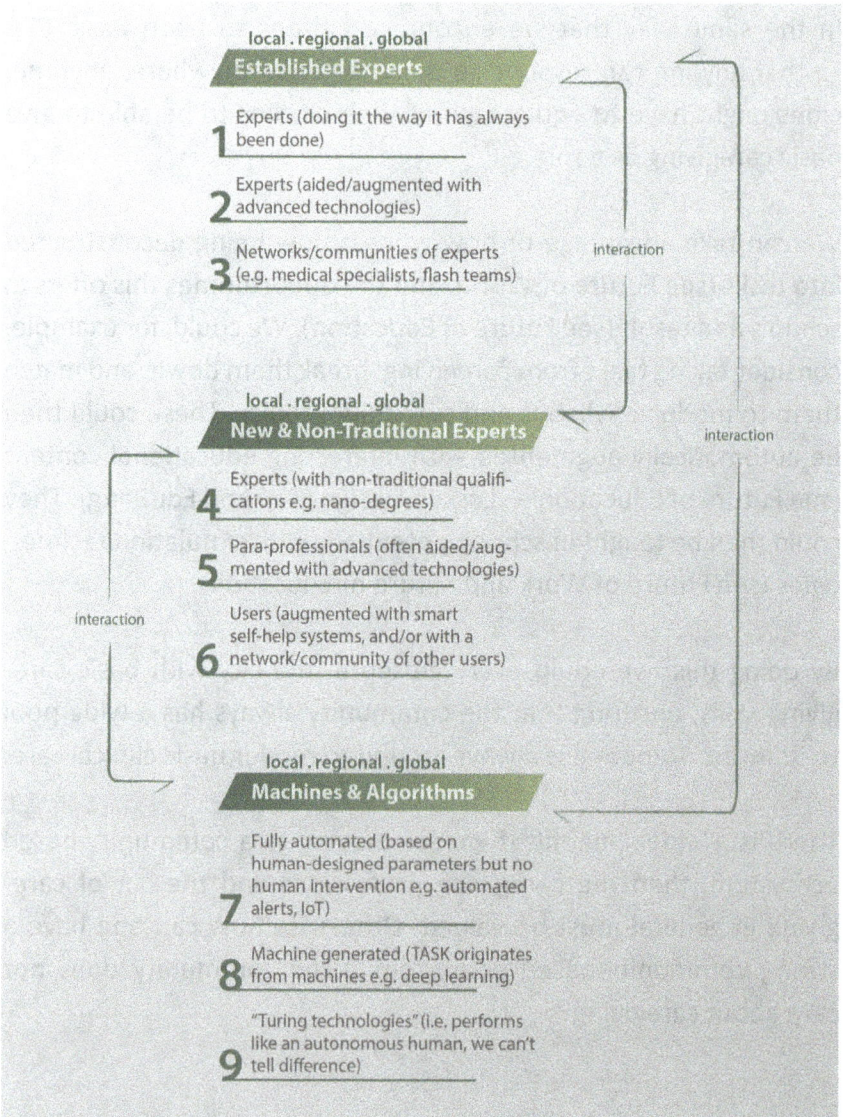

**local . regional . global**
**Established Experts**

1 Experts (doing it the way it has always been done)

2 Experts (aided/augmented with advanced technologies)

3 Networks/communities of experts (e.g. medical specialists, flash teams)

interaction

**local . regional . global**
**New & Non-Traditional Experts**

interaction

4 Experts (with non-traditional qualifications e.g. nano-degrees)

5 Para-professionals (often aided/augmented with advanced technologies)

6 Users (augmented with smart self-help systems, and/or with a network/community of other users)

interaction

**local . regional . global**
**Machines & Algorithms**

7 Fully automated (based on human-designed parameters but no human intervention e.g. automated alerts, IoT)

8 Machine generated (TASK originates from machines e.g. deep learning)

9 "Turing technologies" (i.e. performs like an autonomous human, we can't tell difference)

Fig 6.2.5 — Healthcare Expertise Will be Deconstructed, Democratised and Diversified

# Elevate Interactions

In the future of healthcare, when we empower, energise, and equip people, community and society with skills and values, we have effectively created new pockets of expertise across the city. In our section on Future of Work, we called this the democratisation, deconstruction and diversification of TASK (technologies, attitudes, skills and knowledge — see Fig 6.2.5)

There are now more sources of expertise in the future of healthcare. And these new sources of expertise will be constantly interacting with the existing and established experts and professions. It means that we now have to pay more attention to how they interact with each other too.

**6) Elevate the professional-patient relationship: make it an adult-adult partnership**
A doctor in family medicine shared that today's doctor-patient relationship is more like an adult-child relationship, with the doctor prescribing to the patient what needs to be done. In some cases, it is an adult-defiant child relationship: some patients come in armed with lots of information they have found themselves, and seek to challenge the doctors at every turn.

This relationship between an empowered professional and an empowered patient is a new phenomenon that will become increasingly prevalent.

We can do better. We can elevate the relationship of any healthcare professional with a patient or person to an adult-adult partnership. Technologies empower both professionals and people. Professionals have deep experience, and can access the latest medical findings more easily because of that. People can become more activated and literate, and can engage the professionals to become even more so. Taken together, professionals and people can do a lot more for each other.

Besides encouraging them to elevate their relationship, we should also explore and research how we can help them to do so. It could be part of the training outlined in Recommendation 5. Or there could be behavioural nudges and social network incentives. There might even be new technologies and collaborative platforms that could be developed.

It is a new phenomenon. It will become increasingly prevalent. We need to study and understand it.

### 7) Elevate the professional-professional relationship: strengthen — not lessen — how they work and learn together

Digital can be used to strengthen how professionals work and learn from each other. Healthcare professionals often have to work in concert. In the operating theatre for example, there are many roles:

- scrub nurse (e.g., assists surgeon)
- circulating nurse (e.g., paperwork)
- recovery nurse (i.e., post-surgery)
- anaesthesia nurse (e.g., IV insertion)
- technician (e.g., handles equipment)
- assistant (e.g., position patient)
- surgeon

We can strengthen how they work with and learn from each other. Two healthcare professionals we interviewed suggested doing this via simulated and scenario-based team training. For example, at the

Honor Health Medical Simulation Center in USA, team-based "realistic medical training and scenarios" are used. The scenarios can be created by seasoned professionals, thus transferring knowledge from them to others in the team, especially the less experienced or younger medical professionals.

We can strengthen this further through the use of advanced technologies in simulation, scenarios, sensors, robotics and virtual/mixed/augmented reality. They can improve how fast individual and teams of professionals learn, and how well different professionals can work with each other.

We can strengthen how existing professionals work with each other, and with new professionals entering the sector (e.g., data scientists). We must also be careful about using technologies that might weaken these relationships.

We can also extend this to include new types of professionals entering the health sector. For example, one big new group now is in data science and analytics. These professionals are very well versed in Big Data, but they are typically less familiar with health and healthcare. The converse is true: existing healthcare professionals are well versed in healthcare, but often less familiar with data science.

If we leave it to them to only learn from each other when they are working on actual cases together, it will take a long time before a critical mass of professionals who understand each other is built up. It is also risky, given the potential impact on patients' conditions. Putting both new and existing professionals through simulation training (as described earlier) could speed this up. We can scale the number of such professionals more rapidly.

The importance of getting different professionals to work with each other will grow rapidly in the decades ahead. As healthcare systems evolve into health ecosystems and community- or home-based care, we are likely to see even more disciplines becoming a part of health and vice-versa. We explored this with the Nurture/ Ne Zha Bib, which demonstrates how the humble baby bib could become an advanced digital device drawing on multi-disciplines, from genomics to materials to education to health (see section 6.3 — Artefact from the Future).

"Advanced technologies can transform even the traditional bib".

"The bib can fold upwards and play peek-a-boo with the child when the care givers are busy".

"The bib can also expose the child to an expanded vocabulary across different languages as part of early childhood development".

Fig 6.2.6 — Nurture/哪吒 (Ne Zha) Bib (for expanded view and pictures – see section 6.3 — Artefacts from the Future)

Strengthening how well they learn from and work with each other is only one half of elevating the professional-professional relationship. We must also be careful that technologies do not weaken the relationship.[122]

X-rays illuminate this point very clearly. In the past, before patient X-rays were digitised, doctors used to crowd around a lightbox to discuss together what they could see. This was an excellent way to share expertise, help younger doctors learn, and improve the diagnosis. Now that X-rays are digitised, they are now sent directly into the email inboxes of the doctors involved. A doctor we interviewed was concerned if we might have inadvertently lost some of that mutual learning of the past.

In choosing which technologies we use, we need to consider such losses i.e., potential weakening of how professionals learn and work. We do not do this enough. In an interview about how technologies are assessed within the organisation, we were told that a multi-disciplinary team is tasked to do so, applying ROI calculations and cost-benefit analysis. The interviewee was surprised when we asked if they considered what might be lost, and after a pause, shared that they typically do not, acknowledging they should consider doing so in future.

### 8) Elevate the people/peer/professional-machine relationship: integrated interaction
Our healthcare interviewees frequently cited instances where technology solutions were not well integrated with how they or their organisational processes worked. For example, healthcare professionals might prefer to batch similar tasks across many patients together, but the machines they are using require that

they do all the tasks for one patient before they can move on to the next. Another experienced healthcare professional shared that his first question to technology vendors is whether they know the hospital processes and workflows.

We will need advances in how man and machines work with each other. Otherwise, how well (or badly) people, peers, and professionals work with different technologies might become an impediment.

As the technologies used continue to grow, and as many more in the community are involved in the future of healthcare, how well people, peers, and professionals work with different machines and technologies might in itself become an impediment. Stanford University highlights for example that in AI, "[a]dvances in how intelligent machines interact naturally with caregivers, patients, and patients' families are crucial."[123]

We will need to learn to design these technologies to work seamlessly with people, peers and professionals across multiple dimensions. They have to work well within individual tasks (e.g., positioning a needle in surgery), organisations (hospitals, clinics etc.), communities, and homes (e.g., smart homes in future). They also have to work across all of them, from the seamless sharing of data and information across IT systems across organisations, communities, and homes, to the seamless communication and sharing of expertise and experience across people, peers, and professionals.

## Conclusion: Creating Social and Economic Opportunities

To be a successful community-based ecosystem, technologies, infrastructure, interactions, networks, and communities will all have to be built. They create both social and economic opportunities for the public, private, and people sectors. Each sector in the ecosystem collaborates with each other. Each sector strengthens each other. Each sector helps each other innovate and become better.

And the entire ecosystem can be both local and global. We can access and assess the world's needs, resources, and collective intelligence. It could attract the best ideas, innovations and technologies worldwide. The ideas, innovations, and technologies developed within the ecosystem could also travel back out to the world.

The ecosystem serves our health and healthcare needs first and foremost. It has a public and social purpose. Do that well and — intriguingly — it could also create an innovation and economic ecosystem. The ecosystem can strengthen the social capital and economic future of the city.

# 6.3 Healthcare in 2040

## Santa Social

*In your fingers, in your toes, and everywhere you go: health is all around.*

Cities and companies called it "24/7 Care", "365 Shield", and "360 Degrees Protection". The citizens, with their characteristic wry humour, simply called it "The Santa".

And why not?

***"He sees you when you're sleeping/He knows when you're awake."*** The Electronic Health Records had become the Electronic Everything Records. It was not just when you slept or when you woke up, or worked out; round the clock, SuperSuits and sensors could take every imaginable personal, behavioural, and environmental measurement of you.

***"He knows when you've been bad or good."*** All this health data and information were then crunched together with the latest research findings. It would predict instantly if that dreary management meeting you just attended, or that extra piece of fried chicken wing at lunch, or that funky smell in the autonomous public car had heightened your health risks. In short, it told you if you had been "naughty or nice" to yourself.

***"So be good for goodness sake."*** With these predictions, it became much easier to tailor advice, actions, nudges, and treatments to individuals. At the right value and costs. And all at the right time. Many became more motivated, taking preventive and pre-emptive measures before it was too late. Why be bad to yourself when it had become so easy to be good?

Empowering citizens this way had been critical to the healthcare system's evolution into a community-based health ecosystem. Just as critical was that the system had to go social too.[124]

Family and friends chipped in. So did schools, offices, factories, eateries, shopping malls, transport companies, and many other organisations. People, peers, and professionals developed new ways of working with each other. They all understood that each of them would be healthier if all of them were healthier together. Their personal health depended on social and public health.

Costs were thus shared and distributed across multiple segments of the population.[125] Costs were also better matched: the cost savings from preventive health could be redirected to providing high quality acute care for chronic conditions and humane end-of-life for an ageing population.[126]

An intriguing idea arose of a Health Provident Fund. It combined "The Santa" with the social. If everyone agreed to let their Electronic Everything Records be aggregated, they would create an Electronic Everybody's Everything Repository. This Electronic Everybody's Record was a data gold mine. Cities could use it for population

> Empowering citizens and going social were critical to a community-based health ecosystem. This gave rise to an intriguing idea of the Health Provident Fund, that could be a financial buffer for healthcare expenses, retirement and rainy days.

health segmentation. Scientists could use it for research. Companies could develop precision interventions. And all were prepared to pay for the data.

If the right privacy, confidentiality, security, regulations, policies, and practices could be put in place, together with the right business models, every citizen could be paid regularly for their round-the-clock health data.[127] If untouched and compounded over the years, each citizen's Health Provident Fund payments could be a financial buffer for health and healthcare expenses, retirement, and rainy days.

Health could finally become (a little) wealth for everyone.

# Artefact from the Future — SuperSuit/SuperCare/Empathy Suit

**Jaz's Grandparents,**

80s

Digital Peer Tutored (while recuperating in hospital)

## Year 2030: SuperSuit/SuperCare/Empathy Suit[128–152]

*The ultimate wearable: nanofibres, sensors, 4D printing, and industrial design.*

They were all so dedicated. The doctors, the nurses, and the allied professionals. And they never wavered when global pandemics threatened. They stood firm as a fortress, and together with citizens and companies, kept the city safe.

The city was thus thankful that they had resisted taking automation to its logical extreme during the era of Big Automation. It had been so tempting. A lot of spending was needed — ageing populations, increasing clinical complexities, growing specialisations, technology investments, and new technology-overuse syndromes (like the iHunch) — and every day, there were so many vendors knocking on the hospitals' doors promising substantial cost savings. It all sounded so very attractive.

Fortunately, the city and the hospitals decided that not everything that counts can be counted and costed. They decided they had to be smart with people too, and not just with technology. People after all were at the heart of the healthcare system. That is why so many healthcare professionals had joined and stayed in the sector. Technology was not just for cutting costs; it was also to help healthcare professionals and patients.

Technology such as the SuperSuit.

The SuperSuit as a SuperNurse uniform was electrospun to strengthen support for the nurses' backs. Patients were getting heavier and nurses were getting older too. Robots could do a lot of the lifting and carrying of patients, but there remained other activities that required nurses to assist the patients. The SuperNurse ensured that nurses did not injure their backs as they did this.

The SuperSuit as a SuperCare hospital gown was also smart.

SuperCare could be tailored to the patient's personal health and cultural needs. It gave the patients maximum physical and psychological comfort. It sensed the patients' conditions continuously – but not intrusively – and alerted doctors, nurses, and allied professionals if something seemed to be amiss. In fact, SuperCare seemed to do such a good job of knowing how patients felt, and helping patients and professionals at just the right time that it was also dubbed the "Empathy Suit".

The SuperCare hospital gown (@Empathy Suit) was one of the many patient-centric innovations implemented over the decades. Nobody wanted to be hospitalised. But if they were, the experience need not be any more unpleasant than it needed to be. In fact, many patients felt it was rather pleasant. There was so much care, compassion, comfort, and consideration for their needs. It felt warm, safe, and healing. And thanks to the online and in-person classes taught by volunteers, some patients even picked up new technologies, skills, and knowledge, or rekindled childhood loves and hobbies.

There was one remaining persistent complaint: food.

No matter how hard the hospitals tried, the patients always had a bone to pick with the meals provided. The hospitals had even procured the iYummy, a computationally creative super deep learning 4D food fabricator. iYummy crunched a global database of culinary recipes, and cooked up — using locally sourced organic ingredients — a tasty and healthy meal that matched the patient's favourite street foods and dietary requirements. The meals were supposed to be of Michelin-starred standards, but the patients still had a beef with them. It was just not yummy enough. The citizens' passion for their food was insatiable.

Some things were simply beyond even the most advanced technologies. Even in the most advanced healthcare systems.

# SuperCare Hospital Gown

**Additive Manufacturing:** Intelligent Textile fabricated with coated or spun nano materials (i.e. nanofabric) and 4D Printing technology for maximum comfort, support (for weak and injured areas), and style (for confidence). The gown can also be styled to be aspirational (e.g. kids role-playing superheroes) and nostalgic (e.g. an adult's favorite dress when they were younger).

**Sensing & Actuating:** The gown senses and monitors physiological signs of the wearer. Based on these signs, the gown will respond to provide maximum comfort and support to the wearer. The smart fabric in the gown also provides the power (e.g. solar) for these functions.

**Communications & Empathy:** The change in colour of the entire gown in an emergency alerts those around to provide immediate assistance; it also sends calls for help to the nearest healthcare professionals. To replace the conventional drips and lines attached to the body, the gown could provide medication and treatment to the patient by releasing drugs or providing therapy. Parts of the gown can be highlighted to indicate where the patient is experiencing pain. The entire experience of the wearer can also be fed into future training scenarios for healthcare professionals to develop skills and empathy.

**Intelligent Textile Technology.**
Earbuds are printed with the gown to accurately track vital signs such as heart rate, temperature, and respiration rate and to provide aural comfort to the wearer.

**Sense and Alert**.
Colour-change on nano spun earbud cable around patient's neck to display level of urgency for attention needed from healthcare professionals.

**4D-Printed** area allows sleeves to form several programmed shapes to portray confidence from role-play. It could boost morale of a child by responding when the gown senses wearer's anxiety.

**Smart Access** allows intuitive 'peel' to undress/wear gown giving patients full coverage, and great convenience to healthcare professionals (enabled by 4D Printing technology).

**Nanofabric Technology** maintains and adjusts temperature according to wearer's physiological status. It prevents bacterial growth and eliminates moisture and odour. It also prevents static electricity charge.

FEELING FIT

FEELING SUPER

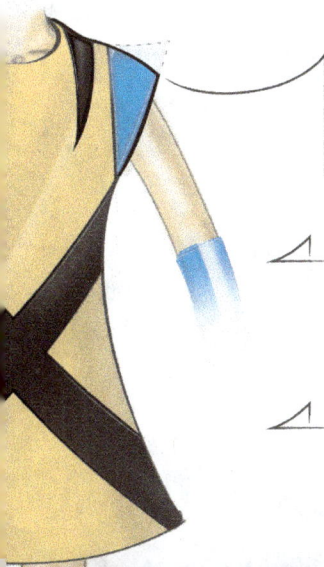

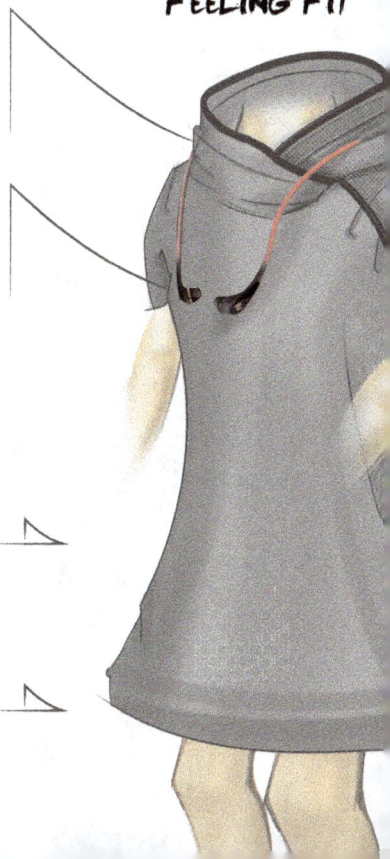

# FEELING FESTIVE

# FEELING NOSTALGI

**Sense and Alert.**
Colour-change on patterns around the collar (or entire gown for Feeling Nostalgic) to signal emergency.

Pattern size changes to signal range of physiological changes.

**Smart Access** allows intuitive 'peel' to undress/wear gown giving patients full coverage, and great convenience to healthcare professionals. (enabled by 4D printing technology)

**Nanofabric Technology**
maintains and adjusts temperature according to wearer's physiological status. It prevents bacterial growth and eliminates moisture and odour. It also prevents static electricity charge.

Fig 6.3.1 — SuperCare/Empathy Suit/SuperSuit

# Artefact from the Future:
# Nurture/哪吒 (Ne Zha) Bib

**Jaz,** 30s,
**and her child**
Investor, Designer,
and User of Bib

## Year 2040: Nurture/哪吒 (Ne Zha) Bib[153–179]
*Aiding childhood development with AI, genomics, sensors, 4D printing, and industrial design.*

Tutor of a vast vocabulary. Protector of marital harmony. Exemplar of gene-environment interaction. And a cute and acute absorber of drool.

The humble bib had been thoroughly transformed by technology.

The company behind it was a joint venture between Asian investors. They saw it as a pan-Asian company that would tap on the best talents and technologies across Asia. They would then take their products into global markets.

The Nurture/哪吒 (Ne Zha) Bib was a worldwide success.

Fig 6.3.2 — Nurture/哪吒 (Ne Zha) Bib Concept Sketches

Parents everywhere loved it. And it was not just because it was the quickest drying bib in the market. Or that it was super green — made of dye-sensitised solar textile, it was powered by the sun.

It was because it was highly affordable (thanks to the decades of manufacturing prowess in Asia), and helped them tackle some of the early childhood developmental challenges, especially for lower-income families.

It could expose young children to multiple languages. This built the child's facility with two or more languages. One would have imagined this would become less important now that universal translators were ubiquitous. But in a globalised world, understanding and accessing another culture — literature, arts, commerce — in its own language was more highly prized than ever.

It could measure the children's drool and other vital physiological signs. These alerted parents early if something was not well in their children's health. They could then take preventive action as needed.

Parents also loved the bib because it seemed their children loved it too. It was like a toy. It talked to the children. It played peek-a-boo with them. More subtly, their children could point to the designs they liked, and the bib would be 4D-printed. Parents could even transform their children's imperfect crayon stick figure drawings onto the bib, and that would be the one printed.

There was a group of parents who felt especially grateful for the bib. They were the parents whose children had been determined to have a risk allele (different forms of a gene found at the same location on a chromosome) for major depression. Research had shown that the genetic effect depended on how stressful the environment was for the baby. The bib reduced the possibility that any severe arguments the parents had amongst themselves — arising from their own stressful lives — would spill over into a "stressor" for their children. The bib shielded their children, making the effect of the risk allele irrelevant. It also reminded the parents visually to cool down.

Fig 6.3.3 — Nurture/哪吒 (Ne Zha) Bib

The bib's phenomenal success had taken a long time, lots of patient capital from the investors, and many successive generations of design.

The first set of technologies it incorporated were the AI and chatbot technologies that had become mainstream by the early 2020s. These first made it possible for the bib to detect harsh words spoken by the parents, and transform them into nurturing words the child would hear. They then made it possible for the bib to become interactive. The bib could understand and respond to young children's imperfect pronunciation (and sometimes gibberish). Then came the universal translators for multi-lingual exposure.

The second set of technologies were the advances in manufacturing and materials. 4D printing — and more broadly additive manufacturing — of fabrics, and spinning of nanofibres, made it possible to customise both the shape and properties of the printed product. Parents could simply go online and pick the features and functions they wanted.

The most interesting set of technologies had to do with genomic sequencing. Parents were beginning to come round to the benefits of sequencing the genome of their children to catch potential problems early. It was also becoming cheaper and easier to do so. More importantly, the efficacy had improved with advances such as Genome-Wide Association Studies.

It was in this last set of technologies that Singapore played a special part in the Nurture/哪吒 (Ne Zha) Bib's success. Because of its decades-long history of building up the biomedical sector and its focus on design, a cluster of genomic industrial designers evolved. They were designers who aspired not to change the genome, but the environment. They understood that for selected genetic effects, the magnitude of the effects depended on the environment. The bib was validation of their design philosophy.

For many then, be it parenting, investment, research, or design, Ne Zha was a godsend (pun intended).

Oh, and one more thing. The best part about the bib? It was machine washable too.

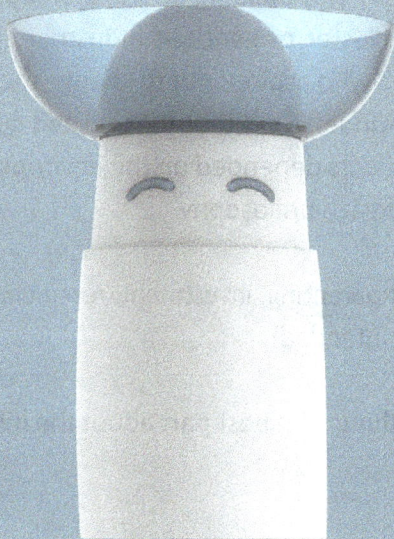

# Nostalgia.

More than a health monitor.

More than a security blanket.

Jazper is a time machine that bridges generations.

Through Jazper you can connect to our venerable team of digital peer tutors, who can teach your child the languages, lingo, dialects, songs, stories, nursery rhymes, and games from days of yore.

Be it Dayung Sampan, 点虫虫, நான் ஆணையிட்டால், or 쎄쎄쎄, you can introduce your child to whole new worlds.

JAZPER

comfort. companion. culture

*Dayung Sampan - a Malay folk song; 点虫虫(Dim Chung Chung) - a Cantonese nursery rhyme; நான் ஆணையிட்டால்(Naan Anai Ittál) - a classic Tamil movie song; 쎄쎄쎄(Ssae Ssae Ssae) - hand clap game

# 7   Conclusion: The Scalable City

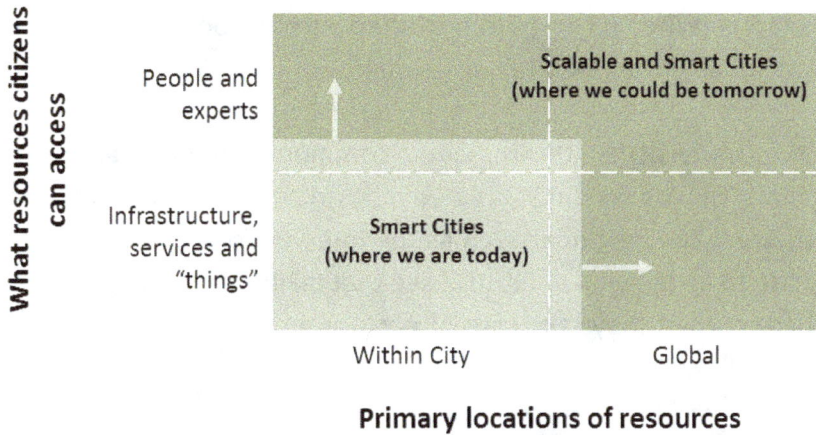

Fig 7.1 — Scalable and Smart Cities

Because of digital, cities are now scalable.

In the future of work, cities can deconstruct and reconstruct work for every citizen, across the city, and around the world. No task is too small or no collection of tasks is too big. Cities can always reconfigure the right mix of local, regional, and global expertise, across individuals or networks, and across established, new, and machine-based expertise.

*Living Digital 2040* discovered that the future of work, education, and healthcare is also a scalable one. Scalability of this digital nature holds the promise that small cities could once again transcend their small physical size. Imagine if in the digital decades ahead, digital does for data, information, tasks, and expertise, what the shipping container did for physical goods?

In the future of education, personalised learning and teaching can be scaled. Each and every child or teacher can teach the world. Each and every child or teacher can also be guided by a global team of experts. When it comes to nurturing personal peaks, no interest is too small and no ambition is too big.

In the future of healthcare, cities' community-based health eco-systems can be as small as needed, or as big as desired. The ecosystem can be global even as the interventions are local, even precise to an individual. Citizens see that taking care of themselves is better when they take care of each other, locally, regionally, or globally.

Scalability of this nature is possible because digital technologies have made possible what used to be "expensive, impossible, or often simply never imagined."[1] Professor Michael Batty from the University College London states that new information technolo-gies have transformed the costs of interaction so much that they have put the "role of networks, communications, and interactions to the fore in thinking about the contemporary and future city."[1] Consequently, what a city chooses to do for the future "is doomed to failure if it does not account for networks that span the globe."[1]

We have to go beyond the physical. He states that his recent book *The New Science of Cities* "outlines a science no longer exclusively based on theories of location... [and] we must underpin our theo-ries with ideas about how we relate to each other... [and] switch our traditional focus from locations to interactions."[2]

According to him, "by the end of this century, the physical form of cities in terms of their locations is likely to be massively different…[and] this change is likely to be so great that we have no idea as to what the city of the medium-term future will look like."[3]

*Living Digital 2040* was our attempt to imagine what that future might look like. We explored the future of work, education, and healthcare, spheres which we experience intimately and where many of our interactions and connections are forged.

In doing so, we were also discovering what a Scalable City might look like.

Becoming a Scalable City is of tremendous import to small cities like Singapore. It holds the promise that the small city-state could once again transcend its small size.

In the pre-digital decades in the last century, the shipping container made it possible to break down and reassemble physical goods. This transformed global shipping, trading, and entire social and economic systems.[4]

What if, in the decades ahead, digital does for data, information, tasks, and expertise, what the shipping container did for physical goods? What if it disrupted the laws of urban scaling?[5] We might see an equally wide-ranging transformation across our global, social, and economic systems. Transformations that we had assumed to be expensive or impossible, and opportunities we had simply failed to or yet to imagine.

And what if we built on what we are already doing in smart cities?[6] Many smart city strategies focus on what goes on within the city.

It is digital, but bounded by the physical. *Living Digital 2040* shows that we can build on the physical but scale digitally to reap massive, widespread benefits.

We might come to a new understanding of how to make ourselves relevant to citizens, companies, and cities worldwide.[4] We might be able to do more individually and internationally. We could transform cities, even as digital transforms us.

We could become smart and scalable.

# Annex A

## Research Methods

We adopted a qualitative approach for this project, meaning that we choose to focus on the interpretations and the meanings that people attach to their words and worldviews. We wanted depth, definition, and diversity on how people live their lives according to the ways they make meaning of their environments and actions, including how they use and avoid technology.

**We wanted depth, definition, and diversity on how people live their lives according to the ways they make meaning of their environments and actions, including how they use and avoid technology.**

We used in-depth interviews, participant observation, and group discussions; these are established data collection methods in qualitative research on organisations and work,[1][2][3] education,[4][5] and healthcare.[6][7][8][9] We also purposefully consulted empirical studies in those areas with a technology bent.

As a result, we did not use quantitative research methods such as surveys and probability sampling. The use of structured survey questionnaires with constrained response options (such as "yes/no" and a range of sentiments from "extremely agree" to "extremely disagree" toward a set of opinions) and statistically representative sample of a target population will not collect the appropriate types of data that help us achieve our research objectives.

As highlighted earlier, we focus on three domains, namely healthcare, education, and work. We chose these three domains as they affect all of us in how we live, work, and play, and also because each can be considered a social institution that everyone will experience first-hand at various points in his or her life.

We are aware that these three domains are discrete and separate from but also closely connected to one another. Therefore, we took the effort to make sure that during our interviews, while focusing on specific domains and sub-areas, we also introduced and connected different domains to one another at various junctures.

For example, we asked respondents identified for inquiry in the work domain to also reflect on their university or college lives, especially how their courses could be refined to better match their aspirations and envisioned career trajectories, and whether any change should be introduced at all. This line of inquiry during our fieldwork made sure that we connected work and education.

**We used a combination of purposive and snowball sampling methods where we actively sought individuals with specific characteristics and sets of experiences so that we could achieve a satisfactory level of variety and depth in our data.**

We used a combination of purposive and snowball sampling methods where we actively sought individuals with specific characteristics and sets of experience so that we achieve a satisfactory level of variety and depth in our data. To illustrate using the education domain, we met educators who had experience in implementing successful, innovative programs using technology in their schools. We also spoke broadly with educators in leadership positions, such as principals, as well as "rank and file" teachers. Insights from such "thick" data were instrumental in building rich and empirically-grounded scenarios, personas, and artefacts.

We also alternated between the various respondent groups and domains during interviews to ensure that our emergent understanding holds across contexts, and to purposefully seek out re-orienting or disconfirming observations from other groups.[10] At the same time, we tempered this approach with practical contingencies. For example, we gave priority to respondents with challenging schedules (e.g., senior management) and those who were exiting their organisations.

> **This approach of focusing on diversity and depth of ideas gels well with our future-oriented focus because we want to discover and elaborate on elements of scenarios that are insufficiently or not yet imagined.**

We took into account the frequency of an issue or topic being raised during fieldwork and analysis, and paid close attention to novel, less obvious, and counterintuitive and contrarian views that challenged received or conventional wisdom or majority opinions that we had inadvertently taken for granted. For example, one respondent reminded us that ageing will more broad-based than we realised; not only are patients going to be older, but also our nursing professionals and home caretakers. They too would need help to perform repetitive menial tasks, such as helping patients to shower and turning patients over in bed to prevent them from developing bed sores.

This approach of focusing on diversity and depth of ideas gels well with our future-oriented focus because we want to discover and elaborate on elements of scenarios that are insufficiently or not yet imagined. It gave us an opportunity to think more clearly and created psychologically safe spaces to raise contrarian viewpoints. In total, we spoke with 174 individuals, over the course of 71 interviews and nine workshops. The table below provides a summary of the profiles of and the context under which we interacted with the 174 individuals.

| Fieldwork type | Number of Individuals | Remarks |
| --- | --- | --- |
| Interview respondents: Experts and decision makers | 41 | Examples are scholars, senior executives in firms and healthcare organisations (e.g., hospitals), and school principals. |
| Interview respondents: Users and technology managers | 40 | Examples are teachers, heads of department, students, doctors, nurses, as well as junior associates and managers in firms. |
| Workshops (n = 9) | 93 | This number includes participants at one workshop (n = 14) that was not organised by the Living with Technology project team, but where observation notes of the discussion were taken by project members who sat in during the event.\
\
Workshops included topics such as GCE "A" levels economics lesson plan prototyping, future of healthcare and wearables, and future of education, among others. |
| Grand Total | 174 | |

Fig A1 — Summary of Fieldwork

**This distinguished group of individuals also provided us a more balanced view of decision-making processes, both ground-up and top-down, in existing hierarchies.**

In addition, we had extensive conversations with another 20 individuals — whom we called our "informants" — across a broad spectrum of contexts and for various purposes, and often more than once. For example, some informants were individuals involved in similar futures-oriented initiatives like us or were domain subject experts. Others served as our gatekeepers who facilitated access to decision makers in organisations such as firms, schools, and hospitals. Given the control of access and their busy schedules, we would not have been able to reach several decision makers without our informants. This distinguished group of individuals also provided us with a more balanced view of decision-making processes, both ground-up and top-down, in existing hierarchies. They provided us with invaluable background information and expedited our learning process. We wrote memos of our interactions with them and those documents added depth and definition to our analysis, scenarios, personas, and artefacts.

# Fieldwork

We started fieldwork after we received approval from SUTD's Institutional Review Board in early August 2015. Our fieldwork lasted about 14 months, from August 2015 to October 2016.

We conducted most interviews one-to-one, but for some respondents, we conducted group interviews when we knew in advance that the individuals involved preferred familiar faces and a less threatening environment to discuss novel and less mainstream ideas.[11][12] Recognising the limits of recall biases in retrospective accounts, we actively corroborated the events, issues, and themes noted in these interviews with archival data analysis and if possible, field observations in-situ (e.g., on school premises). Where appropriate, we also cued the interviewees on key events that they had omitted, or challenged them on inconsistencies that we identified. On average, the one-to-one and group interviews lasted about 75 minutes, with the shortest interview lasting for 45 minutes and the longest, for about 160 minutes.

**Recognising the limits of recall biases in retrospective accounts, we actively corroborated the events, issues, and themes noted in these interviews with archival data analysis and if possible, field observations in-situ (e.g., school premises).**

During interviews, we used a semi-structured interview guide. We typically started with broad questions, trying to understand the background of our interview respondents and participants from our group discussions and workshops. As the discussion proceeded, we focused on the key domain areas (i.e., healthcare, education, and work) and their "function" (as subject matter experts, users and practitioners, and decision makers).

We audio-recorded our interviews when respondents gave us explicit permission to do so. During interviewing, we discontinued or suspended recording when respondents requested us to do so or when we sensed that recording was hampering candid and deep discussion. We also did not request for or perform any recording when we knew in advance that respondents were not receptive to audio recording.

**We worked as a team, taking on multiple roles that were contingent on field situations.**

We worked as a team, taking on multiple roles that were contingent on field situations. We adopted the stance of "known" investigators in the field;[13] our respondents, participants, and informants/gatekeepers understood our presence was as researchers who sought to understand their work. One presentational tactic that we deployed was "acceptable incompetence" to draw out elaboration especially from our respondents.[14] This quintessential student role encouraged them to flesh out their arguments, such as relating real-life examples to illustrate their abstract ideas. On other occasions, we displayed "selective competence"[15] to enhance our credibility as informed members of the community. For example, in discussing the future of healthcare with some healthcare professionals, we used appropriate professional jargon (e.g., "patient activation") during our discussions.

We also treated our semi-structured interview guide as a dynamic document that would undergo iterative revisions to better extract and organise data collection.[16] The approach resembled that of a "guided" conversation rather than one that followed the strict protocol of a survey questionnaire.

We took notes during the guided conversations. Those notes formed the basis for more reflective memos that captured overall impressions, key ideas, analytical and methodological points from the interviews, as well as queries to follow up on during fieldwork.

**We held bi-weekly team progress updates, which provided opportunities for other members to play devil's advocate by asking clarification questions or offering alternative interpretations and gave us the occasion to clarify our interpretations of our qualitative data.**

Some members actively questioned while others actively took notes. Those who took notes did not interrupt the discussions. On some occasions, we deliberately deferred our impulse to immediately pen down respondents' animated interactions on interesting topics until they moved on to more mundane ones. This strategic delay between observation and documentation reduced the likelihood of eliciting negative responses from our informants, particularly when they were disclosing sensitive information.

When it was impossible to take physical notes, we made an effort to take "head notes" to tag significant events.[17] Following recommended ethnographic practices to mitigate the effects of memory loss,[18] we attempted to type field notes within the day. Interview and discussion memos/notes and interview transcripts (when available) were shared with every member in the research team.

Besides noting what had transpired in the field, we also surfaced tentative themes, and identified technical queries and methodological concerns that we needed to address or anticipate in subsequent fieldwork.

As much as possible, we ensured that our methods adhered to the standards and guidelines articulated in established ethnographic/ qualitative fieldwork in social sciences as well as other related disciplines such as information systems, organisation and management studies.[19][20][21]

We held bi-weekly team progress updates, which provided opportunities for other members to play devil's advocate by asking clarification questions or offering alternative interpretations and gave us the occasion to clarify our interpretations of our qualitative data. In essence, we ensured that the team did not conclude our analysis prematurely.[22]

## Analysis and Building Scenarios, Personas, and Artefacts

We captured our initial impression of our field notes, memos, and interview transcripts as early as possible, even as fieldwork was still on-going, to prevent ourselves from becoming de-sensitised to the data.[23]

During our early readings of our data and materials, we focused on bracketing "codable moments".[24] This refers to fragments of text in our field notes, memos, interview transcripts, and other materials. These "codable quotes" provided the foundation for subsequent first and second cycle coding.[25] For example, we bracketed how our respondents talked about technology and the extent to which they felt it would help or hinder their jobs and lives.

During first cycle coding, we used a hybrid of simultaneous coding (also known as double coding), descriptive coding, in vivo coding, process coding ("-ing" coding), and initial coding (also known as open coding) to carefully identify those highlighted instances.

During second cycle coding, we focused on conceptual scaffolding.[26] We moved between both stages to refine the codes so that while they remained close to the data, they also became more precise and abstract in terms of what we wanted to label, and how we planned to organise them.

We used this coding strategy to identify and refine what digital technologies can do for us (e.g., "live, love, learn, and earn," "see, sense, experience and empathise" in Drivers of Change) and dimensions of scenarios (e.g., struggle versus strive in Future of Work).

The various codes and concepts generated also provided the foundation on which we built our scenarios, personas, and artefacts. The interim scenarios, personas, and artefacts (see Fig A2 for the sketches of patient SuperSuits) gave us concrete materials to focus our discussions and refine the narratives that connect the various elements in our report in a coherent and purposive fashion. It was through such discussions and continual refinement that we came to create a family of personas, instead of a loose collection of disparate and unconnected individual personas that are more common in other future or scenario planning initiatives.

Using a family of personas helped us to not only articulate and elaborate familial ties — which can be considered the most basic and closest social relationships and interactions in society — but also provided a means for us to examine the various life stages (e.g., being in school, entering work force, retiring) and the process of ageing, as well as to weave them into our narratives. These narratives are critical in our analytical and creative process as they depict the aspirations, fears, and needs of our personas, and flesh out the social contexts inhabited by our personas. This in turn informs our

Using a family of personas helped us to not only articulate and elaborate familial ties-which can be considered the most basic and closest social relationships and interactions in society- but also provided a means for us to examine the various life stages (e.g., being in school, entering work force, retiring) and the process of ageing, as well as to weave them into our narratives.

scenarios and the features of our artefacts.[27][28] Our treatment of the narratives and their relationship to personas and artefacts closely resembles how "design fiction" is used in some research communities (e.g., Human-Computer Interaction(HCI)) to establish a "problem space" where ideas, however weak or equivocal they might be, could be explored, refined, or even discarded.[29]

Done this way, we are confident that the scenarios, personas, and artefacts that we curate and create are not only outcomes of "disciplined imagination",[30] but also careful abstractions of properties and traits that were based on authentic lived experience.

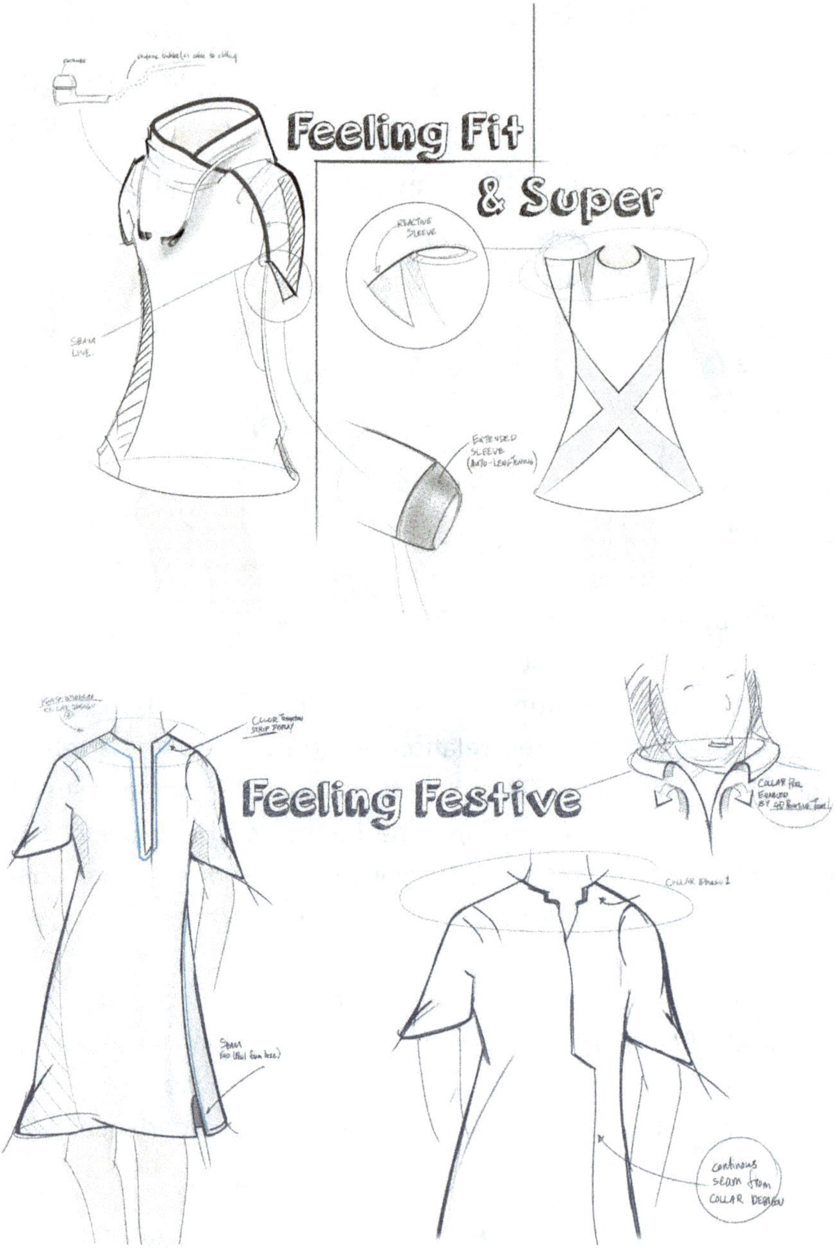

Fig A2

# Annex B

## Consolidated View of Selected Major Technology Advances and Forecasts

| Area | By 2020 | By 2025 | By 2030 | Source(s) |
|---|---|---|---|---|
| Autonomous Navigation | Autonomous vehicles will be capable of driving in any modern town or city with clearly lit and marked roads and demonstrate safe driving comparable to a human driver. Performance of autonomous vehicles will be superior to that exhibited by human drivers in such tasks as navigating through an industrial mining area or construction zone, backing into a loading dock, parallel parking, and emergency braking and stopping. | Autonomous vehicles will be capable of driving in any city and on unpaved roads, and exhibit limited capability for off-road environment that humans drive in, and will be as safe as the average human-driven car. Vehicles will be able to safely cope with unanticipated behaviours exhibited by other vehicles. Vehicles will also be able to tow other broken down vehicles. Vehicles will be able to reach a safe state in the event of sensor failures. | Autonomous vehicles will be capable of driving in any environment in which humans can drive. Their driving skills will be indistinguishable from humans except that robot drivers will be safer and more predictable than a human driver with less than one year's driving experience. Vehicles will be able to learn on their own how to drive in previously unseen scenarios. | Roadmap for US Robotics[1] |
| Human-like Dexterous Manipulation | 1) Low-complexity hands with small numbers of independent joints will be capable of robust whole-hand grasp acquisition. 2) Robotic manipulators for surgery should be able to perform snake-like manoeuvres at great depth. Manipulators for everyday objects should be expanded to handle more general objects and tasks (pick up, deliver, turn know, open door, push button, move slider, etc.) 3) Snake-like robotic instruments will enable surgeons to perform simple natural orifice translumenal endoscopic surgical procedures in the abdomen via the stomach. Robot assistants will aid healthcare workers in safely moving patients in and out of hospital beds. | 1) Medium-complexity hands with ten or more numbers of independent joints and novel mechanisms and actuators will be capable of whole-hand grasp acquisition and limited dexterous manipulation. 2) Micro-scale robots should be able to assist in dexterous microsurgery in small structures such as the eye, as well as cellular-scale surgery. Mobile manipulation with on-board power and computation should manipulate objects in everyday environments safety. 3) Snake-like surgical robots will be capable of high-dexterity surgical tasks throughout the body and should also be miniaturised to enable precise microsurgical repairs. Tetherless centimetre-scale robots will be introduced that can perform interventional tasks inside the body such as removing polyps or modulating blood flow. Assistive robots will interact with impaired individuals to perform self-care tasks, such as grooming, hygiene, and dressing. | 1) High-complexity hands with tactile array densities, approaching that of humans and with superior dynamic performance, will be capable of robots whole-hand grasp acquisition and dexterous manipulation of objects found in manufacturing environments used by human workers. 2) Groups of tetherless millimetre- and micron-scale robots will be able to both swim through bodily fluids and bore through tissue to perform highly localised therapies. Assistive robots will autonomously perform general care-related tasks in human-centric environments with only high-level supervision. | Roadmap for US Robotics[2] Research Roadmap for Medical & Healthcare Robotics[3] |

| Area | By 2020 | By 2025 | By 2030 | Source(s) |
|---|---|---|---|---|
| Edutainment | Automated understanding of emotional & physiological state: Wireless wearable sensors to detect and classify, as well as to some degree predict user physiological state. | Robots educating and entertaining humans; multi-modal communication - assessment of a person's emotional state and physical expression of emotions and gestures; challenge is to produce robots with sufficient functionality to generate novelty and fascination and maintain the interest of a person over a significant time span at a suitable price. | N.A. | Roadmap for US Robotics[4] Europe Robotics Roadmap 2010[5] |
| Automated Understanding of Human State and Behaviour During Robot Interaction | 1) Context-appropriate guidance: A robot should be able to recognise and classify human behaviour and intent - in a modified environment and/or people are augmented to make perception easier. 2) Robots will be able to have the ability to capture human state and behaviour (aided with wearable sensors) in controlled environments (e.g., physical therapy sessions, doctor's offices) with known structure and expected nature of interactions. Data from such sessions will begin to be used to develop models of the user that are useful for developing general schemes for optimising robot interactions. | Robots will be able to automatically classify human state and behaviour from lightly instrumented users (lightweight sensors), in less structured settings (e.g., doctor's offices and homes with less-known structure), visualise those data for the user and the healthcare provider, and choose appropriate interactions for individual users based on the classification. | Robots will be able to detect, classify, predict, and provide coaching for human activity within a known broad context (e.g., exercise, office work, dressing) with minimal use of obtrusive sensors. The robot will be able to provide intuitively visualised data for each user, based on the user's needs. Decisions for robot interactions based on the ongoing classification of state and behaviour will use algorithms validated as effective in rigorous experimental studies. | Roadmap for US Robotics[6] |
| Secure and Safe Robot Behaviour | We will exhibit cost-effective, inherently safe actuation, and light-weight/high-strength robot bodies in surgical and socially assistive robotics for in-clinic and in-home testing for specific tasks. We will fully characterise and theoretically counter security vulnerabilities present in remotely controlled surgical devices. | We will create affordable and safe standardised translational research platforms (both hardware and software) for safe in-clinic and in-home robot evaluation with heterogeneous users (healthcare providers, family, patient). We will collect longitudinal data on safety and usability. We will test secure communication links suitable for secure telemedical interventions. | We will achieve safe deployment of robot systems in unstructured environments (e.g., homes, outdoor settings) involving human-machine interaction in real-time with unknown users, with minimal training and using intuitive interfaces. We will deploy proven safe surgical robots with partial autonomy capabilities which achieve greater precision than human surgeons. We will develop secure telemedical interventions mediated by the open Internet. | Roadmap for US Robotics[7] |
| Sustainable and Manufacturing | Manufacturing process will recycle 10% of raw materials, reuse 50% of the equipment, and use only 90% of the energy used in 2010 for the same process. | Manufacturing process will recycle 25% of raw materials, reuse 75% of the equipment, and use only 50% of the energy used in 2010 for the same process. | Manufacturing process will recycle 75% of raw materials, reuse 90% of equipment, and use only 10% of the energy used in 2010 for the same process. 2040: Spare capacity built into supply chains for resilience; Products remanufactured and redesigned with recovery in mind; Tougher environmental standards for products; New ways of 'pricing' the environment; Business models based on reuse and remanufacturing and services; 2050+: Products kept in 'production loop' where they are remanufactured to original/better specs | Roadmap for US Robotics[8] Roadmap for UK Manufacturing[9] |

| Area | By 2020 | By 2025 | By 2030 | Source(s) |
|---|---|---|---|---|
| Nano-manufacturing | N.A. | N.A. | Nano-manufacturing for nano-robots for drug delivery, therapeutics, and diagnostics. | Roadmap for US Robotics[10] |
| Safe Robot | Assist humans in performing services useful to well-being of human and equipment, e.g., sophisticated therapy for stroke patients | N.A. | N.A. | Europe Robotics Roadmap 2010[11] |
| Surveillance and Intervention | Automated understanding of emotional & physiological state: Wireless wearable sensors to detect and classify, as well as to some degree predict user physiological state. | Use of flying robotic platforms for surveillance will increase, and in the long term accomplish more complex tasks like responding to sudden and unexpected events and identifying abnormal activities or potentially dangerous situations. | 1) Automated understanding of emotional and physiological state: off-the-shelf wireless physiologic sensing devices to inter-operate with computer- and robot-based coaching systems for real-time bio-feedback and classification of user physiological and emotional state. 2) Robotic systems able to provide full suite of physical feedback to human operator. A surgeon or caregiver should be able to feel forces, detailed surface textures, and other physical properties of a remote patient. 3) Quantitative diagnosis & assessment: connect and access bioelectrical signals with wearable or implantable devices. Quantitative analysis of data to inform in situ diagnosis as well as long-term patient tracking. 4) Achieve semi-automated and automated surgical assistants that use fully real-time image-to-model generation (including geometry, mechanics, and physiological sensing). | Research Roadmap for Medical & Healthcare Robotics[12]  Europe Robotics Roadmap 2010[13] |
| Cooking | N.A. | Demonstration of a robot where programming by demonstration can be used for complex task learning such as meal preparation in a regular home. | N.A. | Roadmap for US Robotics[14] |

| Area | By 2020 | By 2025 | By 2030 | Source(s) |
|---|---|---|---|---|
| Quantitative Diagnosis | N.A. | N.A. | We can accomplish connecting and easily accessing biophysical signals with wearable or implantable devices in real time. This is linked to integrated unencumbered multimodal sensing and intuitive data visualisation environment for the user and caregiver. Real-time algorithms enable not only off-line but also online quantitative analysis of such data to inform in situ diagnosis as well as long-term patient tracking or skill acquisition. Systems are developed for in-home use and detection of early symptoms of pervasive disorders, such as autism spectrum disorder, from behavioural data. Similarly, the progression of degenerative motor disorders, such as Parkinson's or muscular dystrophy, can be monitored. Finally, by closing the loop, adaptive training algorithms based on the quantitative assessments will enable personalised protocols for procedural training for surgeons, or individualised rehabilitation regimens for cognitive and sensorimotor impairments. | Roadmap for US Robotics[15] |
| Health Data / Sensor-based | We will establish common and open infrastructure for data collection, building on recently developed models such as ROSbag, but extended to a broader class of medical robots and devices. | We will create sharable data sets for key research areas, including robotic surgery, prosthetics, rehabilitation, and in-home living. | We will create cloud-based analysis frameworks, with baseline performance of existing algorithms, to enable rapid design, development, and evaluation cycles for medical robotics research. | Roadmap for US Robotics (Healthcare & Medical Robotics)[16] |
| Telemedicine | N.A. | We will test secure communication links suitable for secure telemedical interventions. | We will develop secure telemedical interventions mediated by the open Internet. | Roadmap for US Robotics[17] |
| Indoor GPS | Accurate indoor positioning systems for mobile manipulators, particularly in dynamic environments | N.A. | N.A. | |

| Area | By 2020 | By 2025 | By 2030 | Source(s) |
|---|---|---|---|---|
| **Mobility and Manipulation** | Robots exploit diverse mobility mechanisms in research laboratories to navigate safely and robustly in unstructured 2D environments and perform simple pick and place tasks. Relevant objects are either from a very limited set or possess specific properties. Robots create semantic maps about their environment through exploration and physical interaction but also through instruction from humans. They are able to reason about tasks of moderate complexity, such as removing obstructions, opening cabinets, etc. to obtain access to other objects. | Given an approximate and possibly incomplete model of the static part of the environment (possibly given a priori or obtained from databases via the Internet, etc.), service robots are able to reliably plan and execute a task-directed motion in service of a mobility or manipulation task. The robot builds a deep understanding of the environment from perception, physical interaction, and instruction. The robot navigates multi-floor environments through stairways. The robot modifies its environment to increase the chances of achieving its task (e.g., remove obstructions, clear obstacles, turn on lights), and detects and recovers from some failures. | Service robots including multiple mobility mechanisms such as legs, tracks, and wheels perform high-speed, collision-free, mobile manipulation in completely novel, unstructured, and dynamic environments. They perceive their environment, translate their perceptions into appropriate, possibly task-specific local and global/short- and long-term environmental representations (semantic maps), and use them to continuously plan for the achievement of global task objectives. They respond robustly to dynamic changes in the environment (e.g., unexpected perturbation due to being pushed or jostled). They are able to interleave exploratory behaviour when necessary with task-directed behaviour. They interact with their environment and are able to modify it in intelligent ways so as to ensure and facilitate task completion. This includes reasoning about physical properties of interactions (sliding, pushing, throwing, etc.) between the robot, objects it comes into contact with, and the static parts of the environment. | Roadmap for US Robotics[18] |
| **Cognition and Skills Acquisition** | 1) Demonstration of a robot that can learn skills from a person through gesture and speech interaction. In addition, acquisition of models of a non-modelled indoor environment. 2) Robots can learn a variety of basic skills through observation, trial and error, and from demonstration. These skills can be applied successfully under conditions that vary slightly from the ones under which the skill was learned. Robots can autonomously perform minor adaptations of acquired skills to adapt them to perceived differences from the original setting. | 1) A robot that interacts with users to acquire sequences of new skills to perform complex assembly or actions. The robot has facilities for recovery from simple errors encountered. 2) As perceptual capabilities improve, robots can acquire more complex skills and differentiate specific situations in which skills are appropriate. Multiple skills can be combined into more complex skills autonomously. The robot is able to identify and reason about the type of situation in which skills may be applied successfully. The robot has a sufficient understanding of the factors that affect the success so as to direct the planning process in such a way that chances of success are maximised. | 1) A companion robot that can assist in a variety of service tasks through adaptation of skills to assist the user. The interaction is based on recognition of human intent and re-planning to assist the operator. 2) The robot continuously acquires new skills and improves the effectiveness of known skills. It can acquire skill-independent knowledge that permits the transfer of single skills across different tasks and different situations and the transfer of skills to novel tasks. The robot is able to identify patterns of generalisation for the parameterisation of single skills and across skills. | Roadmap for US Robotics[19] |

# References

## Chapter 1

1) Negroponte, N. (2014, March). *A 30-year history of the future.* Retrieved from https://www.ted.com/talks/nicholas_negroponte_a_30_year_history_of_the_future where 1980 was identified as the year he first conceived of this point.

2) Transcript of Prime Minister Lee Hsien Loong speech at Smart Nation launch on 24 November 2014. Retrieved October 17, 2016, from http://www.pmo.gov.sg/mediacentre/transcript-prime-minister-lee-hsien-loongs-speech-smart-nation-launch-24-november

"Therefore our vision is for Singapore to be a Smart Nation – A nation where people live meaningful and fulfilled lives, enabled seamlessly by technology, offering exciting opportunities for all. We should see it in our daily living where networks of sensors and smart devices enable us to live sustainably and comfortably. We should see it in our communities where technology will enable more people to connect to one another more easily and intensely. We should see it in our future where we can create possibilities for ourselves beyond what we imagined possible."

3) MIT openDOOR (2003, August). Interview with: Professor Sherry Turkle. (2003, August). Retrieved October 17, 2016, from http://web.mit.edu/sturkle/www/pdfsforstwebpage/ST_Open%20Door%20interview.pdf

4) Manyika, J., Lund, S., Bughin, J., Woetzel, J., Stamenov, K., & Dhingra, D. (2016, February). *Digital globalization: The new era of global flows.* Retrieved from http://www.mckinsey.com/business-functions/digital-mckinsey/our-insights/digital-globalization-the-new-era-of-global-flows

5) Glasmeier, A., & Christopherson, S. (2015). *Thinking about smart cities. Cambridge Journal of Regions, Economy and Society, 8*(1), 3-12. doi:10.1093/cjres/rsu034

6) Jacobs, J. (1964). *The death and life of great American cities (Vintage Books Edition, December 1992)*. London: Pelican Books.

7) Batty, M. (2013). *The new science of cities*. Cambridge, MA: MIT Press.

Professor Batty also reiterated this in his talk and conversations at the Lee Kuan Yew Centre for Innovative Cities at the Singapore University of Technology on 31st March 2016.

8) Sassen, S. (2012). *Cities in a world economy (4th ed.)*. Los Angeles, CA: Sage.

9) Fukuda-Parr, S. (2001). Making new technologies work for human development. *United Nations Development Programme, Human Development Report*. Retrieved from http://hdr.undp.org/en/content/human-development-report-2001

10) Goldin, I., & Kutarna, C. (2016). A*ge of discovery: Navigating the risks and rewards of our new renaissance*. New York, NY: St. Martin's Press.

From p. 77:
"Education is both a consequence of development - something people who can choose, do choose - and a catalyst for further health and income gains..."

From p. 79:
"Article 26 of the 1948 Universal Declaration of Human Rights... affirms that 'Everyone has the right to education' and 'Education shall be directed to the full development of the human personality.'"

11) Lee, K. Y. (2006). Good governance and the wealth of nations. Speech presented at Singapore 2006 Annual Meetings of the Boards of Governors of the International Monetary Fund (IMF) and World Bank Group in Singapore. Retrieved October 17, 2016, from http://www.mas. gov.sg/annual_reports/annual20062007/42_TEAM.htm

Excerpt (from podcast that was available for download then): "...we watched the developed world and said what is it they have which we don't have? Good infrastructure. Highly educated workforce that knows how to use that infrastructure."

12) Rosling, H. (2006, June). Hans Rosling: The best stats you've ever seen [Video file]. Retrieved from http://www.ted.com/talks/hans_ rosling_shows_the_best_stats_you_ve_ever_seen
Transcript retrieved October 17, 2016, from http://www.ted.com/talks/ hans_rosling_shows_the_best_stats_you_ve_ever_seen/transcript

"The countries are moving more or less in the same rate as money and health, but it seems you can move much faster if you are healthy first than if you are wealthy first... but health cannot be bought at the super-market. You have to invest in health. You have to get kids into schooling. You have to train health staff. You have to educate the population..."

13) The Economist Intelligence Unit (2016). High aspirations, stark realities: Digitising government in South-east Asia. Retrieved October 17, 2016, from https://www.eiuperspectives.economist.com/sites/default/ files/EIU_Microsoft%20DigitisingGov_briefing%20paper_Jan2016.pdf

Health, education and finance are deemed as the top 3 areas where cloud computing will have the greatest impact both now and in 3 years.

14) Acemoglu, D. (2012). *The world our grandchildren will inherit: The rights revolution and beyond.* (NBER Working Paper No. 17994). Cambridge, MA: National Bureau of Economic Research. doi:10.3386/ w17994

15) Topol, E. J. (2015). *The patient will see you now: The future of medicine is in your hands*. New York, NY: Basic Books.

16) Ford, M. (2015). *Rise of the robots: Technology and the threat of a jobless future*. New York, NY: Basic Books.

17) Brynjolfsson, E., & McAfee, A. (2014). *The second machine age: Work, progress, and prosperity in a time of brilliant technologies*. New York, NY: WW Norton & Company.

18) Wagner, T., & Compton, R. A. (2012). *Creating innovators: The making of young people who will change the world*. New York, NY: Scribner.

19) Stanford University (2016, September). Artificial Intelligence and Life in 2030. In *One Hundred Year Study on Artificial Intelligence: Report of the 2015-2016 Study Panel*. Stanford, CA: Stanford University. Retrieved October 17, 2016, from https://ai100.stanford.edu/2016-report

Not coincidentally, Stanford University's September 2016 report for the "One Hundred Year Study on Artificial Intelligence (AI100)" is not only titled "Artificial Intelligence and Life in 2030", it also chooses to focus on what life will be like in North American cities, and the "specific changes affecting the everyday lives of the millions of people who inhabit them." The areas it chose included employment, education and healthcare, in addition to the domains of transport, low-resource communities, public safety and security, service robots, and entertainment.

The report also suggests why the impact of technology should be analysed according to specific domains. They write "[t]hough drawing from a common source of research, each domain reflects different AI influences and challenges…"

# Chapter 2

1) Pinker, S. (2015). *The sense of style: The thinking person's guide to writing in the 21st century*. New York, NY: Penguin Books.

2) MacGregor, N. (2010). Episode 1 - Mummy of Hornedjitef [Radio programme transcript]. BBC. Retrieved October 17, 2016, from http://www.bbc.co.uk/programmes/articles/1Ryz8NYSgfT4Llx7rHQ1hPf/episode-transcript-episode-1-mummy-of-hornedjitef

3) Schwartz, P. (1996). *The art of the long view: Paths to strategic insight for yourself and your company*. New York, NY: Doubleday.

4) United Kingdom, HM Government. (n.d.). The futures toolkit: Tools for strategic futures for policy-makers and analysts. Retrieved from https://www.gov.uk/government/uploads/system/uploads/attachment_data/file/328069/Futures_Toolkit_beta.pdf

5) Dunne, A., & Raby, F. (2013). *Speculative everything: Design, fiction, and social dreaming*. Cambridge, MA: MIT Press.

6) Tetlock, P. E., & Gardner, D. (2016). *Superforecasting: The art and science of prediction*. New York, NY: Broadway Books.

7) Saffo, P. (2007). Six rules for effective forecasting. *Harvard Business Review, 85*(7/8), 122-131.

8) Personas. (n.d.). Retrieved October 17, 2016, from http://www.usability.gov/how-to-and-tools/methods/personas.html

9) Nesta. (n.d.).Future Londoners. Retrieved October 17, 2016, from http://www.nesta.org.uk/news/future-londoners

10) Norman, D. (2004, November 16). Ad-Hoc Personas & Empathetic Focus. Retrieved October 17, 2016 from http://www.jnd.org/dn.mss/personas_empath.html

11) Institute for the Future (n.d.) Artifacts from the Future. Retrieved October 17, 2016, from http://www.iftf.org/what-we-do/foresight-tools/artifacts-from-the-future/

Silicon Valley-based Institute for the Future writes "Imagine that you could take an archaeologist's expedition to the future to collect objects and fragments of text or photos to understand what daily life will be like in 10, 20, or 50 years. Artifacts from the Future give us this tangible experience of the future. They make the details of a scenario concrete, helping us to understand, almost first-hand, what it will be like to live in a particular future."

12) MacGregor, N. (2011). *A history of the world in 100 objects*. New York, NY: Viking.

13) Hon, A. (2013). *A history of the future in 100 objects*. Amazon Digital Services.

# Chapter 3

1) Diamandis, P. H., & Kotler, S. (2012). *Abundance: The future is better than you think*. New York, NY: Free Press.

2) TheDrawShop. (2014, January 23). Peter Diamandis - Exponential technology [Video file]. Retrieved October 17, 2016, from https://www.youtube.com/watch?v=IaZOux1Qqwg

3) Perez, C. (1983). Structural change and assimilation of new technologies in the economic and social systems. *Futures, 15*(5), 357-375.

4) The Economist. (1999, February 18). Innovation in industry - Catch the wave. Retrieved October 17, 2016, from http://www.economist.com/node/186628

5) Moody, J. B., & Nogrady, B. (2010). *The sixth wave: How to succeed in a resource-limited world*. Sydney: Vintage Books.

6) Schwab, K. (2016). The fourth industrial revolution. Geneva: World economic forum. Retrieved October 17, 2016, from http://www3.weforum.org/docs/Media/KSC_4IR.pdf

7) George, B. J., & Devarajan, V. (2016, June 27). Industry 4.0: Business in the age of personalisation. Retrieved October 17, 2016, from https://www.weforum.org/agenda/2016/06/industry-4-0-business-in-the-age-of-personalisation

8) The Economist (2012, April 21). The third industrial revolution. Retrieved October 18, 2016, from http://www.economist.com/node/21553017

9) Rifkin, J. (2011). *The third industrial revolution: How lateral power is transforming energy, the economy, and the world*. New York, NY: Palgrave Macmillan.

10) Brynjolfsson, E., & McAfee, A. (2014). *The second machine age: Work, progress, and prosperity in a time of brilliant technologies*. New York, NY: W. W. Norton & Company.

11) United Nations Educational, Scientific and Cultural Organization. (n.d.). What are Open Educational Resources (OERs)? Retrieved October 17, 2016, from http://www.unesco.org/new/en/communication-and-information/access-to-knowledge/open-educational-resources/what-are-open-educational-re-sources-oers/

12) DARPA | Cyber Grand Challenge. (n.d.). Retrieved October 17, 2016, from http://archive.darpa.mil/grandchallenge/

13) Davidson, L. (2016, July 3). Using Facebook to transfer money? That could be a reality sooner than you think. Retrieved October 17, 2016, from http://www.telegraph.co.uk/business/2016/07/03/using-facebook-to-transfer-money-that-could-be-a-reality-sooner/ The social media scheme is part of a plan to build "a system for the future with capabilities to store multiple [proxy IDs]"

14) Dutta, N. (2016, July 20). Adaptive hearables - JWT intelligence. Retrieved October 17, 2016, from https://www.jwtintelligence.com/2016/07/adaptive-hearables/

15) Stanford University (2016, September). Artificial intelligence and life in 2030. In *One hundred year study on artificial intelligence: Report of the 2015-2016 study panel*. Stanford, CA: Stanford University. Retrieved October 17, 2016, from https://ai100.stanford.edu/2016-report

16) Gibney, E. (2016). Chinese satellite is one giant step for the quantum internet. *Nature, 535*, 478–479. doi:10.1038/535478a

The craft that launched in August is first in a wave of planned quantum space experiments.

17) Claburn, T. (2016, January 31). Google AI can spot image location with 'superhuman' accuracy. Retrieved October 19, 2016, from http://www.informationweek.com/big-data/big-data-analytics/google-ai-can-spot-image-location-with-superhuman-accuracy-/d/d-id/1324489

18) The World Bank (n.d.). Identification for development. Retrieved October 19, 2016, from http://www.worldbank.org/en/programs/id4d

"Identification is core to development because it is a key enabler for... Financial Inclusion: Accessible, secure, and verifiable ID systems can help expand the use of financial services by approximately 375 million unbanked adults in developing countries."

19) Whittaker, Z. (2016, August 7). 'Quadrooter' flaws affect over 900 million Android phones. Retrieved October 19, 2016, from http://www.zdnet.com/article/quadrooter-security-flaws-affect-over-900-million-android-phones/

20) Simmons, A. C. (2016, June 13). Artificial intelligence produces realistic sounds that fool humans. Retrieved October 19, 2016, from http://news.mit.edu/2016/artificial-intelligence-produces-realistic-sounds-0613

"Video-trained system from MIT's Computer Science and Artificial Intelligence Lab could help robots understand how objects interact with the world... For robots to navigate the world, they need to be able to make reasonable assumptions about their surroundings and what might happen during a sequence of events. One way that humans come to learn these things is through sound. For infants, poking and prodding objects is not just fun; some studies suggest that it's actually how they develop an intuitive theory of physics. Could it be that we can get machines to learn the same way? Researchers from MIT's Computer Science and Artificial Intelligence Laboratory (CSAIL) have demonstrated an algorithm that has effectively learned how to predict sound... When shown a silent video clip of an object being hit, the algorithm can produce a sound for the hit that is realistic enough to fool human viewers. This "Turing Test for sound" represents much more than just a clever computer trick: Researchers envision future versions of similar algorithms being used to automatically produce sound effects for movies and TV shows, as well as to help robots better understand objects' properties."

21) Ho, O. (2015, April 29). Restoring sights and sounds from past for future generations. The Straits Times. Retrieved October 19, 2016, from http://news.asiaone.com/news/singapore/restoring-sights-and-sounds-past-future-generations

22) Turkle, S. (2012). *Alone together: Why we expect more from technology and less from each other*. New York, NY: Basic Books.

23) Boyd, D. (2014). *It's complicated: The social lives of networked teens*. New Haven, CT: Yale University Press.

24) Topol, E. J. (2015). *The patient will see you now: The future of medicine is in your hands*. New York, NY: Basic Books.

25) Weinswig, D. (2016, July 7). Are you smarter than my clothing? Retrieved October 17, 2016, from http://www.forbes.com/sites/deborahweinswig/2016/07/07/are-you-smarter-than-my-clothing-2/#3217bb937598

"The idea is to add an additional function to textiles," Juan Hinestroza, Associate Professor of Fiber Science at Cornell, told Syracuse's News Channel 9. "Our ultimate goal is to have fibers that can hear, fibers that can see, fibers that can sense, fibers that can act, and your clothing becomes like a second skin."

26) Christensen, C. M. (1997). *The innovator's dilemma: When new technologies cause great firms to fail*. Boston, MA: Harvard Business School Press.

Disruptive change is also often typically associated with radical innovation. Due to the long term time frame (~25 years) of our report, we assume that disruptive change can also be associated with incremental innovation. Incremental innovations that build on each other over years and decades eventually tip over, causing disruption.

Recent examples of this include the personal computing, gaming, and mobile communications.

This can also be seen in the case studies in Harvard professor Clayton Christensen's *The Innovator's Dilemma*:

| Entrant Disruptive Technology | Year of Entry into Market | Year Entrant Disruptive Technology Met Existing Market's Needs | Number of Years Taken |
|---|---|---|---|
| 8-inch rigid disk drive | 1978 | 1988 | 10 years |
| 5.25 inch rigid disk drive | 1980 | 1987 | 7 years |
| 3.5 inch rigid disk drive | 1985 | 1988 | 3 years |

It took at least 3 years for new entrants to disrupt the incumbents. Disruption did not appear silently. It also did not happen suddenly. Our emphasis is thus on the pace of change i.e. whether disruptive change happens rapidly or incrementally.

27) Norman, D. A., & Verganti, R. (2014). Incremental and radical innovation: Design research vs. technology and meaning change. *Design Issues*, *30*(1), 78-96.

"Radical or disruptive change are driven by changes in technology and meaning (such as the purpose of the technology or innovation), thus offering a different proposition."

28) Pratt, G. A. (2015). Is a Cambrian explosion coming for robotics?. *The Journal of Economic Perspectives*, *29*(3), 51-60.

Rapid and incremental are also dynamic. The field of robotics provides an excellent example of this. Past incremental improvements feed into present accelerating advances, which feed into future potential rapid disruption. Where possible in the table below, we have quoted the text verbatim, except where additional explanations to aid readability are deemed to be necessary.

| Where robotics is today is because of past incremental improvements in: | Why the pace of change is picking up today – accelerating advances in: | What could rapidly disrupt the field – Artificial Intelligence (AI) innovations in |
|---|---|---|
| 1. Computing performance<br>2. Electromechanical design tools and numerically controlled manufacturing tools<br>3. Electrical energy storage<br>4. Electronics power efficiency<br>5. Wireless digital local communications<br>6. Performance and scale of Internet<br>7. Worldwide data storage<br>8. Global computation power | 1. Learning (e.g. through mimicking) from what people do (e.g. through vision systems or videos)<br>2. Deep learning and/or new AI algorithms and techniques that replicate the perceptual parts of the human brain<br>3. High speed sharing and access across robots of each robot's learning<br>4. Simulation – robots can explore and experiment with possible solutions and options | Current deep learning and/or new AI algorithms and techniques only replicate some of the perceptual parts of the human brain.<br><br>The game changer will be when it can replicate more of those perceptual parts, as well as other human cognitive functions like episodic memory and "unsupervised learning" (the clustering of similar experiences without instruction).<br><br>We should also keep in mind that when we say "replicate", it might not mean that we mimic exactly how the human brain does it. It could be a clever combination of technology that achieves the same end point but using different methods. |

29) Susskind, R. E., & Susskind, D. (2016). *The future of the professions: How technology will transform the work of human experts*. New York, NY: Oxford University Press.

Authors interviewed both established firms and new entrants across a spectrum of professions (healthcare, education, consulting, tax, audit, architecture, divinity). They all agreed their professions were being transformed by technology. Where they disagreed was the 'pace of change'. In other words, whether it would be rapid or incremental depends on their view of the future.

30) Nordhaus, W. D. (2015). *Are we approaching an economic singularity? Information technology and the future of economic growth* (NBER Working Paper No. 21547). Cambridge, MA: National Bureau of Economic Research. doi: 10.3386/w21547

Yale Professor William Nordhaus examined if we are approaching a Singularity. Singularity is the point where "rapid growth in computation and artificial intelligence will cross some boundary or Singularity, after which economic growth will accelerate sharply at an ever-increasing pace of improvements and cascade through the economy".

He put this idea through seven economic tests. The tests suggest that we are not approaching a Singularity, and even if we were, it would be at least 100 years from now.

31) Frey, C. B., Osborne, M., Holmes, C., Rahbari, E., Garlick, R., Friedlander, G., ... & Wilkie, M. (2016). *Technology at work v2. 0: the future is not what it used to be*. CitiGroup and University of Oxford.

"The cost of investing in and implementing robotics are coming down, all around the word. For example, 'payback periods for robots... 1.7 years for the most popular auto robot used in China; less than half a year in metal manufacturing in Germany; less than 2 years for hospital co-bots...'); we expect that this trend of falling costs are also happening for other technological advances.

32) Saffo, P. (2007). Six rules for effective forecasting. *Harvard Business Review*, *85*(7/8), 122-131. Retrieved October 19, 2016, from https://hbr.org/2007/07/six-rules-for-effective-forecasting

"As futurist Roy Amara pointed out to me three decades ago, there is a tendency to overestimate the short term and underestimate the long term. Our hopes cause us to conclude that the revolution will arrive overnight. Then, when cold reality fails to conform to our inflated expectations, our disappointment leads us to conclude that the hoped-for revolution will never arrive at all — right before it does."

33) Gartner Hype Cycle. (n.d.). Retrieved October 19, 2016, from http://www.gartner.com/technology/research/methodologies/ hype-cycle.jsp

Technologies take time to reach wide adoption is also echoed by Gartner's Hype Cycle Research Methodology.

34) World Economic Forum. (2015, September). Deep shift: Technology tipping points and societal impact. Retrieved October 19, 2016, from http://www3.weforum.org/docs/WEF_GAC15_Technological_Tipping_Points_report_2015.pdf

35) Martin-Brualla, R., Gallup, D., & Seitz, S. M. (2015). Time-lapse mining from internet photos. *ACM Transactions on Graphics (TOG)*, *34*(4), 62:1-62:8. New York, NY: ACM

36) Karp, H. (2015, March 06). Turning a profit from music mashups. Retrieved October 19, 2016, from http://www.wsj.com/articles/turning-a-profit-from-music-mashups-1425687517

37) Seenit. (n.d.). Retrieved October 19, 2016, from https://seenit.io/

38) Musical Orbit. (n.d.). Lessons with the very best classical and jazz musicians in the world . Retrieved October 19, 2016, from http://www.musicalorbit.com/

39) Retelny, D., Robaszkiewicz, S., To, A., Lasecki, W. S., Patel, J., Rahmati, N., Doshi, T., Valentine, M., & Bernstein, M. S. (2014, October). Expert crowdsourcing with flash teams. In P*roceedings of the 27th Annual ACM Symposium on User Interface Software and Technology* (pp. 75-85). New York, NY: ACM.

40) Andrews, B. (2016). The future of work could be "flash teams". Retrieved October 17, 2016, from https://engineering.stanford.edu/news/melissa-valentine-re-inventing-way-we-work

41) Edmondson, A. C. (2012, April). Crisis management - Teamwork on the fly. Retrieved October 17, 2016, from https://hbr.org/2012/04/teamwork-on-the-fly-2

42) Deng, J., Dong, W., Socher, R., Li, L. J., Li, K., & Fei-Fei, L. (2009, June). Imagenet: A large-scale hierarchical image database. In *IEEE Conference on Computer Vision and Pattern Recognition*, 2009. (pp. 248-255). IEEE.

43) Markoff, J. (2012, November 19). Seeking a Better Way to Find Web Images. Retrieved October 17, 2016, from http://www.nytimes.com/2012/11/20/science/ for-web-images-creating-new-technology-to-seek-and-find.html?_r=1

In an example of breaking down and scaling up, in 2009, the world's largest visual database - ImageNet - was built using Amazon's Mechanical Turk service. In 2012, it had 14 million objects labelled, and over 300 publications had used or cited it for computer vision research. By Sep 2016, the research paper detailing the research had been cited close to 2800 times.

44) O'Connor, S. (2016, June 14). The gig economy is neither 'sharing' nor 'collaborative'. Retrieved October 17, 2016, from https://www.ft.com/content/8273edfe-2c9f-11e6-a18d-a96ab29e3c95

45) Aloisi, A. (2015). Commoditized workers: The rising of on-demand work, a case study research on a set of online platforms and apps. *4th Conference of the Regulating for Decent Work Network*. (July 2015).

Different platforms facilitate the slicing up of different types of jobs. These include:
- Task Rabbit for manual tasks
- MBA & Company, and Eden McCallum for consultants
- TopCoder and Innocentive for programmers, engineers and scientists
- Axiom and Upcounsel for lawyers

46) The Economist. (2015, January 03). The on-demand economy - Workers on tap. Retrieved October 19, 2016, from

http://www.economist.com/news/leaders/21637393-rise-demand-economy-poses-difficult-questions-workers-companies-and

47) O'Connor, S. (2015, October 9). The human cloud: A new world of work. A drive to divvy up and scatter jobs into a virtual world of workers raises questions about the outcome. Retrieved October 19, 2016, from http://www.ft.com/cms/s/2/a4b6e13e-675e-11e5-97d0-1456a776a4f5.html#axzz4J0jv232W

48) Hill, A. (2015, November 25). After 17 Harvard case studies, Haier starts a fresh spin cycle. Retrieved October 19, 2016, from http://www.ft.com/intl/cms/s/0/4afb31b0-91eb-11e5-bd82-c1fb87bef7af.htm

An example of breaking down work into smaller pieces and drawing on global resources is Haier. "In its home country, the group is reinventing itself again as a set of open "entrepreneurial platforms", serving — and served by — hundreds of "microenterprises". Not only will these micro-enterprises compete to design, build and distribute products Haier users say they want, but they will also be able to vie with one another for staff and for capital, from Haier and from outside investors... Its 20 platforms include its "diet ecosystem" (based around smart fridges), its "atmosphere ecosystem" (air conditioners and purifiers) and Goodaymart Logistics, a distribution network that is the key to fulfilling the company's promise that it can deliver anywhere in China within 24 hours.

Goodaymart now operates independently, in partnership with Alibaba, the ecommerce group, distributing goods for Haier's competitors as well as its original parent. It works through some subcontracted "vehicle micro-enterprises" (truck-owners, in other words)... One entrepreneurial team, calling itself iSee Mini, uncovered a market for televisions that projected the image on to the ceiling so pregnant women could watch more comfortably. China Daily recounted recently that another part of the group had supplied a loving son in Hefei with an air-conditioner for

his calligrapher father, embossed with his dad's favourite phrase —
"God Rewards the Diligent"."

49) Udacity. (n.d.). Nanodegree Programs. Retrieved October 23, 2016,
from https://www.udacity.com/nanodegree

50) edX. (n.d.). MicroMasters programs : Advance your career &
accelerate your masters. Retrieved October 23, 2016, from
https://www.edx.org/micromasters

51) The Scientist. (2016, October 1). Thirty years of progress.
Retrieved October 23, 2016, from http://www.the-scientist.com/
?articles.view/articleNo/47150/title/Thirty-Years-of-Progress/

52) Horne, M., Khan, H., & Corrigan, P. (2013, April). People powered
health: health for people, by people and with people. London: Nesta,
2013. Retrieved October 23, 2016, from https://www.nesta.org.uk/sites/
default/files/health_for_people_by_people_and_with_people.pdf

53) The Economist. (2015, September 10). Digital Taylorism. Retrieved
October 18, 2016, from http://www.economist.com/news/business/
21664190-modern-version-scientific-management-threatens-
dehumanise-workplace-digital

54) Acemoglu, D. (2012). *The world our grandchildren will inherit:
the rights revolution and beyond* (NBER Working Paper No. 17994).
Cambridge, MA: National Bureau of Economic Research. doi:10.3386/
w17994

55) Ibid.

# Chapter 4

1) Stanford University (2016, September). Artificial intelligence and life in 2030. In *One hundred year study on artificial intelligence: Report of the 2015-2016 study panel*. Stanford, CA: Stanford University. Retrieved September 1, 2016, from https://ai100.stanford.edu/2016-report

2) Frey, C. B., Osborne, M., Holmes, C., Rahbari, E., Garlick, R., Friedlander, G., ... & Wilkie, M. (2016). Technology at work v2.0: The future is not what it used to be. CitiGroup and University of Oxford.

3) Gordon, Robert J. (2016) *The rise and fall of American growth: The U.S. standard of living since the civil war*. Princeton, NJ: Princeton University Press.

4) Mokyr, J., Vickers, C., & Ziebarth, N. L. (2015). The history of technological anxiety and the future of economic growth: Is this time different?. *The Journal of Economic Perspectives, 29*(3), 31-50.

5) Autor, D. H. (2015). Why are there still so many jobs? The history and future of workplace automation. *The Journal of Economic Perspectives, 29*(3), 3-30.

6) Frey, C. B., & Osborne, M. A. (2013). The future of employment: how susceptible are jobs to computerisation. Retrieved on September 1, 2016 from http://www.oxfordmartin.ox.ac.uk/downloads/academic/The_Future_of_Employment.pdf

7) Frey, C. B., & Osborne, M. (2015). *Technology at work: The future of innovation and employment*. Citi GPS: global perspectives & solutions.

8) Brynjolfsson, E., & McAfee, A. (2014). *The second machine age: Work, progress, and prosperity in a time of brilliant technologies*. New York, NY: W. W. Norton & Company.

9) Susskind, R. E., & Susskind, D. (2016). *The future of the professions: How technology will transform the work of human experts.* New York, NY: Oxford University Press.

10) Ford, M. (2015). *Rise of the robots: Technology and the threat of a jobless future.* New York, NY: Basic Books.

11) Ross, A. (2016). *The Industries of the future.* New York, NY: Simon and Schuster.

12) Marsh, P. (2012). *The new industrial revolution: Consumers, globalization and the end of mass production.* New Haven, CT: Yale University Press.

13) Diamandis, P. (2012, June 28). Abundance – The future is better than you think. Retrieved October 19, 2016, from http://singularityhub.com/2012/06/28/abundance-the-future-is-better-than-you-think/

14) Harvard Business School. (2000, April 12). Done deals: Venture capitalists tell their story: Featured HBS John Doerr. Retrieved October 19, 2016, from http://hbswk.hbs.edu/archive/1799.html

15) The Economist. (2015, January 03). The on-demand economy - Workers on tap. Retrieved October 19, 2016, from http://www.economist.com/news/leaders/21637393-rise-demand-economy-poses-difficult-questions-workers-companies-and

16) The Economist. (2014, December 29). The future of work - There's an app for that. Retrieved October 20, 2016, from http://www.economist.com/news/briefing/21637355-freelance-workers-available-moments-notice-will-reshape-nature-companies-and

17) Lee, K. Y. (2012). *The Singapore story: Memoirs of Lee Kuan Yew.* Singapore: Marshall Cavendish International (Asia).

18) Solow, R. M. (1987, July 12). We'd better watch out. [Review of the book manufacturing matters: The myth of the post-industrial economy by S.S. Cohen & J. Zysman]. Retrieved October 20, 2016, from http://www.standupeconomist.com/pdf/misc/solow-computer-productivity.pdf

19) Shapiro, C., & Varian, H. R. (1999). Information rules: A strategic guide to the network economy. Boston, MA: Harvard Business School Press.

20) Christensen, C. M. (1997). *The innovator's dilemma: When new technologies cause great firms to fail*. Boston, MA: Harvard Business School Press.

21) Nordhaus, W. D. (2015). *Are we approaching an economic singularity? Information technology and the future of economic growth* (NBER Working Paper No. 21547). Cambridge, MA: National Bureau of Economic Research. doi: 10.3386/w21547

22) Keynes, J. M. (1933). Economic possibilities for our grandchildren (1930). Essays in persuasion. Retrieved from https://assets.aspeninstitute.org/content/uploads/files/content/upload/Intro_Session1.pdf

23) Goldin, I., & Kutarna, C. (2016). Age of discovery: Navigating the risks and rewards of our new renaissance. New York, NY: St. Martin's Press.

The Industrial Revolution is the historical example citied because it was the first revolution where extensive information was collected about what happened between technology, work and society. Professors Ian Goldin and Chris Kutarna of Oxford University go further back. They suggest in their book Age of Discovery - Navigating the Risks and Rewards of Our New Renaissance, that our current times echoes the Renaissance in innovation, globalisation and social issues. They write for example that then as now, unskilled segments of workers suffered, globalisation contributed to populism, but over time societies adjust and adapt. They also point out that the way to win is to be smart, organised, risk-taking, intellectually open, forward looking, supportive of the arts, and open to immigration, amongst others.

Divides must be closed, otherwise "the energies of the neglected... [are] wasted, or the talents of the disheartened have been withdrawn." Society must also help those who are displaced by the changes by creating social support systems, and by "deliver[ing] more consistently on the promises the moderate world makes, so that more people feel encouraged, and fewer people feel betrayed, by the age they live in."

24) Moretti, E. (2012). *The new geography of jobs*. New York, NY: Houghton Mifflin Harcourt.

25) Alexopoulos, M. (2011). Read all about it! What happens following an economic shock? *American Economic Review, 101*(4): 1144-79.

26) Maddison, A. (1997). Causal influences on productivity performance 1820–1992: A global perspective. *Journal of Productivity Analysis, 8*(4), 325-359.

27) Banks, D. (2014, September 4). The problem of excess genius. Retrieved October 19, 2016, from http://archive.is/HEFEt

28) Ho, O. (2016, July 14). S'pore's future: Be flexible or be first? *The Straits Times*. Retrieved October 19, 2016, from http://www.straitstimes.com/singapore/health/ spores-future-be-flexible-or-be-first

29) Lee, W. U. (2016, July 14). Digital age means companies can be born global players: Iswaran.

The Business Times. Retrieved October 19, 2016, from
http://www.businesstimes.com.sg/government-economy/
digital-age-means-companies-can-be-born-global-players-iswaran

30) Klingebiel, R., & Joseph, J. (2015, August 11). When first movers
are rewarded, and when they're not. Retrieved October 20, 2016, from
https://hbr.org/2015/08/when-first-movers-are-rewarded-and-
when-theyre-not

31) Klingebiel, R., & Joseph, J. (2016). Entry timing and innovation
strategy in feature phones. *Strategic Management Journal, 37*(6),
1002-1020.

32) Moon, Y. (2010). *Different: Escaping the competitive herd*. New York,
NY: Crown Business.

33) Broughton, P. D. (2016, June 30). How viral upstarts wipe the floor
with conformity. Retrieved October 20, 2016, from https://www.ft.com/
content/fc1ddc32-54d1-11e6-9664-e0bdc13c3bef

34) Frey, C. B., & Osborne, M. A. (2013). The future of employment:
How susceptible are jobs to computerisation. Retrieved on September 1,
2016 from http://www.oxfordmartin.ox.ac.uk/downloads/academic/
The_Future_of_Employment.pdf

35) Schumpeter, J. (1950). *Capitalism, socialism and democracy* (3d ed.)
New York, NY: Harper & Row.

36) Goldin, I., & Kutarna, C. (2016). *Age of discovery: Navigating the risks
and rewards of our new renaissance*. New York, NY: St. Martin's Press.

"Since 1960, global average life expectancy has risen by almost two full
decades - from about fifty-two to seventy-one years. It took 1000 years
to achieve the previous twenty-year improvement (although most of

those gains happened after 1850); this time, it took only fifty. In 1990, only one-third of those who died had passed their seventieth birthday. By 2010, it was closer to one-half, and almost one-quarter of all who died had passed their eightieth. In just two decades, eighty has become the new seventy."

The last Renaissance saw schooling transformed from a relative luxury to a more and more precious commodity, of practical and spiritual necessity to many... In the span of a century, education's place in popular society broadened from a curiosity meant for a few to a means of unlocking the potential of many..."

"We are well on our way to realising this vision, beginning with the foundation of all education: literacy. In 1980, almost half the global population (44 per cent) was illiterate. Today, despite rapid population growth, that share has fallen to just one-sixth. In just over a generation, humanity has added three billion literate brains to its ranks."

37) Piketty, T., & Ganser, L. J. (2014). *Capital in the twenty-first century*. Cambridge, MA: Harvard University Press.

"... it was not the lack of inventive ideas that set the boundaries for economic development, but rather powerful social and economic interests promoting the technological status quo."

38) Lee, M. X. (2016, July 12). Don't neglect social inclusion as cities urbanise. *The Business Times*. Retrieved October 19, 2016, from http://www.businesstimes.com.sg/government-economy/ dont-neglect-social-inclusion-as-cities-urbanise

39) Larson, S. (2014, October 9). Serial: The podcast we've been waiting for. The New Yorker. Retrieved October 20, 2016, from http:// www.newyorker.com/culture/sarah-larson/serial-pod-cast-weve-waiting

40) Green, J., & Green, H. (Directors). (n.d.). Vlogbrothers. Retrieved October 19, 2016, from https://www.youtube.com/user/vlogbrothers

41) Zoella (Director). (n.d.). Zoella. Retrieved October 19, 2016, from https://www.youtube.com/user/zoella280390

42) RSA (n.d.). Basic income. Retrieved October 19, 2016, from https://www.thersa.org/action-and-research/rsa-projects/economy-enterprise-manufacturing-folder/basic-income

43) BBC News. (2016, June 5). Switzerland's voters reject basic income plan. Retrieved October 19, 2016, from http://www.bbc.com/news/world-europe-36454060

44) Thornhill, J., & Atkins, R. (2016, May 26). Universal basic income: Money for nothing - Amid anxiety over technological disruption, is a guaranteed payment from the state the future of welfare? Retrieved October 19, 2016, from https://www.ft.com/content/7c7ba87e-229f-11e6-9d4d-c11776a5124d#axzz4IngM2TqJ

45) Acemoglu, D., & Autor, D. (2011). Skills, tasks and technologies: Implications for employment and earnings. Handbook of Labor Economics, 4, 1043-1171.

46) Autor, D. H. (2013, January). The" task approach" to labor markets: An overview (NBER Working Paper No. 18711). Cambridge, MA: National Bureau of Economic Research. doi: 10.3386/w18711

"The canonical production function found in economic models... features a role for labor and a role for machinery (capital), and, in general, these roles are distinct....what precisely is distinctive about the role of each input is left opaque, and the nature of the interactions among them is highly constrained. In particular, capital is either a complement or a substitute for labor, different types of labor are either complements or substitutes for one another, and these roles are essentially fixed. Changes in the supply of each input can of course affect marginal products, but each factor's "purpose" in the production function is both distinct and static.

These restrictions stem from the fact that the canonical production function implicitly equates two distinct aspects of production. One aspect is which factors are used as inputs, e.g., capital, high skill labor, low skill labor. The other is what services these factors provide. In the canonical setup, a factor's identity and its role in the production function are synonymous. In reality, however, the boundary between "labor tasks" and "capital tasks" in production is permeable and shifting....

For the purposes of this discussion, it is essential to draw a distinction between skills and tasks. A task is a unit of work activity that produces output. A skill is a worker's stock of capabilities for performing various tasks. Workers apply their skills to tasks in exchange for wages. Canonical production functions draw an implicit equivalence between workers' skills and their job tasks, as noted above. Here, we emphasize instead that skills are applied to tasks to produce output — skills do not directly produce output. This distinction is of course inconsequential if workers of a given skill always perform the same set of tasks. It is relevant, however, when the assignment of skills to tasks is subject to change, either because shifts in market prices mandate reallocation of skills to tasks or because the set of tasks demanded in the economy is altered by technological developments, trade, or offshoring. In my view, we are currently in such an era."

47) Autor, D. H. (2016, August 15). The shifts - great and small - in workplace automation [Blog post]. *MIT Sloan Management Review*. Retrieved October 19, 2016, from http://sloanreview.mit.edu/article/the-shifts-great-and-small-in-workplace-automation/

48) Smith, A. (2016, March 10). Public predictions for the future of workforce automation. Retrieved October 19, 2016, from http://www.pewinternet.org/2016/03/10/ public-predictions-for-the-future-of-workforce-automation/

49) Ross, A. (2016). *The industries of the future*. New York, NY: Simon and Schuster.

50) Stanford University (2016, September). Artificial intelligence and life in 2030. In *One hundred year study on artificial intelligence: Report of the 2015-2016 study panel*. Stanford, CA: Stanford University. Retrieved October 19, 2016, from https://ai100.stanford.edu/2016-report

51) Chui, M., Manyika, J., & Miremadi, M. (2016, July). Where machines could replace humans - and where they can't. Retrieved October 19, 2016, from http://www.mckinsey.com/business-functions/ business-technology/our-insights/Where-machines-could-replace-humans-and-where-they-cant-yet

52) Kok, S., & Weel, B. T. (2014). Cities, tasks, and skills. *Journal of Regional Science, 54*(5), 856-892.

The interplay between location and interactions is nuanced and complex. But it is useful to note that if task connectivity explains employment change better than spatial concentration of tasks, this is in fact consistent with the premise in Chapter 1 that we need to look beyond physical locations to interactions. As the authors points out, their "framework relies upon the idea that employment grows when job tasks need to be performed in close vicinity and human interactions are important."

Besides the finding described in the main text, the authors also found that:

- "task connectivity and co-agglomeration strongly correlate (0.63 (0.00))" but when co-agglomeration is used in lieu of task connectivity, the coefficient is insignificant i.e. "spatial concentration seem to be less important at the task level".]
- "spatial concentration of our 41 tasks does not seem to explain employment growth."

In addition, they found:
- "insignificant effect of the [diverse task composition] index on employment growth."
- "estimated coefficient for this [labour pool suitability of tasks] index is insignificant and small."
- "task connectivity of high-skilled workers has a stronger impact on employment growth than task connectivity of low-skilled workers... [and] the connectivity between tasks of medium-skilled workers is only moderately correlated with employment growth [which are also consistent with studies on the polarisation of jobs]"
- "co-efficient of task connectivity is similar for both samples [of male and females]"
- "connectivity of job tasks of young workers has a stronger impact on employment growth than connectivity of jobs tasks of older works. This is line with the findings that older workers perform more 'declining' jobs tasks (Autor and Dorn, 2009; Bosch and Ter Weel, 2013)"

The authors also found that:
- "cities with a relatively highly connected task structure seem to be larger, less specialized and more skilled than cities with lower levels of task connectivity. These cities also seem to employ workers for which social skills are relatively more important.. but much less than human capital and [degree of] specialisation [i.e. not as highly correlated]."
- "connectivity between worker skills [calculated in the same way as task connectivity] does not explain employment growth of cities."

53) Ibid.

54) Ibid.

55) Employability & employment centre - Autism resource centre. (n.d.). Retrieved October 19, 2016, from http://www.autism.org.sg/core-services/e2c

56) Tippins, N. T., & Hilton, M. L. (Eds.). (2010, January). *A database for a changing economy: review of the occupational information network (O\* NET)*. Washington, DC: National Academies Press.

The O\*NET database "includes information on skills, abilities, knowledges, work activities, and interests associated with occupations. This information can be used to facilitate career exploration, vocational counseling, and a variety of human resources functions, such as developing job orders and position descriptions and aligning training with current workplace needs".

It was first built in 1939 as the Dictionary of Occupational Titles (DOT). It is telling that the "first edition of the DOT appeared in 1939, when millions of American were out of work." In its Strategic Plan for 2006-2011, the USA Department of Labor spells out two purposes for the O\*NET:
- To support individuals in making education and training decisions and investments, and
- To support business and community needs for a prepared and globally competitive workforce.

57) Ibid.

58) Susskind, R. E., & Susskind, D. (2016). *The future of the professions: How technology will transform the work of human experts*. New York, NY: Oxford University Press.

59) Smith, A. (2016, March 10). Public predictions for the future of workforce automation. Retrieved from http://www.pewinternet.org/2016/03/10/ public-predictions-for-the-future-of-workforce-automation/

60) Susskind, R. E., & Susskind, D. (2016). *The future of the professions: How technology will transform the work of human experts*. New York, NY: Oxford University Press.

61) Topol, E. J. (2015). The patient will see you now: *The future of medicine is in your hands*. New York, NY: Basic Books.

62) Mokyr, J., Vickers, C., & Ziebarth, N. L. (2015). The history of technological anxiety and the future of economic growth: Is this time different?. *The Journal of Economic Perspectives, 29*(3), 31-50.

63) Autor, D. H. (2015). Why are there still so many jobs? The history and future of workplace automation. *The Journal of Economic Perspectives, 29*(3), 3-30.

64) Frey, C. B., & Osborne, M. A. (2013). The future of employment: how susceptible are jobs to computerisation. Retrieved on September 1, 2016 from http://www.oxfordmartin.ox.ac.uk/downloads/academic/ The_Future_of_Employment.pdf

65) Schwab, K. (2016). The fourth industrial revolution. Geneva: World Economic Forum. Retrieved from http://www3.weforum.org/docs/ Media/KSC_4IR.pdf

66) Brynjolfsson, E., & McAfee, A. (2014). *The second machine age: Work, progress, and prosperity in a time of brilliant technologies*. New York, NY: W. W. Norton & Company.

67) Ford, M. (2015). *Rise of the robots: Technology and the threat of a jobless future*. New York, NY: Basic Books.

68) Ross, A. (2016). *The industries of the future*. New York, NY: Simon and Schuster.

69) Kirman, B., Linehan, C., Lawson, S., & O'Hara, D. (2013, April). CHI and the future robot enslavement of humankind: A retrospective. In *CHI'13 Extended Abstracts on Human Factors in Computing Systems* (pp. 2199-2208). New York, NY: ACM.

70) Autor, D. H. (2016, August 15). The shifts - great and small - in workplace automation [Blog post]. *MIT Sloan Management Review*. Retrieved October 19, 2016, from http://sloanreview.mit.edu/article/the-shifts-great-and-small-in-workplace-automation/

71) Bloom, B. S. (1956). Taxonomy of educational objectives. *Vol. 1: Cognitive domain*. New York, NY: McKay.

72) Autor, D. H. (2016, August 15). The shifts - great and small - In workplace automation. *MIT Sloan Management Review*. Retrieved October 19, 2016, from http://sloanreview.mit.edu/article/the-shifts-great-and-small-in-workplace-automation/

73) CSR [Personal interview]. (2016).

As a CSR practitioner shared with us, CSR should be "360 degrees". CSR is the social conscience of the company, and it is as much about how the company treats it employees and partners, as it is about philanthropy and social work etc.

74) DARPA | Cyber Grand Challenge. (n.d.). Retrieved October 18, 2016, from http://archive.darpa.mil/grandchallenge/

75) Coldewey, D. (2016, August 5). Carnegie Mellon's mayhem AI takes home $2 million from DARPA's Cyber Grand Challenge. Retrieved October 20, 2016, from https://techcrunch.com/2016/08/05/ carnegie-mellons-mayhem-ai-takes-home-2-million-from-darpas-cyber-grand-challenge/

76) Arthur, W. B. (2000, September). *Myths and realities of the high-tech economy.* Talk given at Credit Suisse First Boston Thought Leader Forum (Sep. 10, 2000). Retrieved October 18, 2016, from http://tuvalu.santafe.edu/~wbarthur/Papers/Credit_Suisse_Web.pdf

"Deep craft" is a term used by Professor Brian Arthur of Santa Fe Institute to describe knowledge and applied science fed through a culture of praxis [i.e. practice] that is taken for granted [by the practitioner] and difficult to reproduce... or impossible to transfer.

77) Goldin, I., & Kutarna, C. (2016). *Age of discovery: Navigating the risks and rewards of our new renaissance.* New York, NY: St. Martin's Press.

78) The Economist. (2009, February 16). Business process re-engineering. Retrieved October 20, 2016, from http://www.economist.com/ node/13130298

79) The Economist. (2008, July 28). Downsizing. Retrieved October 20, 2016, from http://www.economist.com/node/11773794

80) The Economist. (2015, September 10). Digital Taylorism. Retrieved October 20, 2016, from http://www.economist.com/news/business/ 21664190-modern-version-scientific-management-threatens-dehumanise-workplace-digital

81) Hammer, M. (1990, August). Reengineering work: Don't automate, obliterate. Retrieved October 20, 2016, from https://hbr.org/1990/07/ reengineering-work-dont-automate-obliterate

82) Davenport, T. H. (1995, October 31). The fad that forgot people. *In Fast Company*. Retrieved October 20, 2016, from https://www.fastcompany.com/26310/fad-forgot-people

83) Davenport, T. H., & Stoddard, D. B. (1994). Reengineering: business change of mythic proportions?. *MIS Quarterly*, 18(2), 121-127.

84) Denning, S. (2010, July 2). What HBR won't say: Why BPR failed [Blog post] Retrieved October 20, 2016, from http://stevedenning.typepad.com/steve_denning/2010/07/what-hbr-wont-say-why-bpr-failed.html

85) Highsmith, J., & Cockburn, A. (2001). Agile software development: The business of innovation. *Computer, 34*(9), 120-127.

86) Sutherland, J., Viktorov, A., Blount, J., & Puntikov, N. (2007, January). Distributed scrum: Agile project management with outsourced development teams. In *Proceeding of 40th Annual Hawaii International Conference on System Sciences*, pp. 274a-274a. IEEE.

87) Sousa, A. F. D. (2013). Business oriented applications for Android platform (Doctoral dissertation, Instituto Politécnico de Leiria).

88) The Economist. (2015, September 10). Digital Taylorism. Retrieved October 20, 2016, from http://www.economist.com/news/business/21664190-modern-version-scientific-management-threatens-dehumanise-workplace-digital

89) Dhar, V. (2016, May 17). When to trust robots with decisions, and when not to. Retrieved October 20, 2016, from https://hbr.org/2016/05/when-to-trust-robots-with-decisions-and-when-not-to

90) Ayres, I. (2007). *Super crunchers: Why thinking-by-numbers is the new way to be smart*. New York, NY: Bantam Books.

91) Lee, C. (2016, March 24). Big data predicts who will be invited for a job interview. Retrieved October 19, 2016, from http://discovery.rsm.nl/articles/detail/ 218-big-data-predicts-who-will-be-invited-for-a-job-interview/

92) Gershgorn, D. (2016, September 6). When artificial intelligence judges a beauty contest, white people win. Retrieved October 19, 2016, from http://qz.com/774588/artificial-intelligence-judged-a-beauty-contest-and-almost-all-the-winners-were-white/

93) Autor, D. H. (2015). *Why are there still so many jobs?* The history and future of workplace automation. *The Journal of Economic Perspectives, 29*(3), 3-30.

94) Autor, D. H. (2013). *The "task approach" to labor markets: An overview* (NBER Working Paper No. 18711). Cambridge, MA: National Bureau of Economic Research. doi: 10.3386/w18711

"If the boundary between 'labor tasks' and 'capital tasks' is fluid, what determines the division of labor — or, more precisely, the allocation of tasks — between these factors? At least two forces are central, one technological, the other economic. On the technological front, the boundary between labor and capital shifts primarily in one direction: capital typically takes over tasks formerly performed by labor; simultaneously, workers are typically assigned novel tasks before they are automated. This sequence of task allocation makes intuitive sense: when a task is unfamiliar or poses unexpected obstacles, workers can often draw on outside knowledge and problem-solving skills to devise work-arounds. By contrast, few machines can improvise. Consequently, automating a task requires attaining a level of mastery beyond what is required for a worker to simply perform the task; it must be codified to the point where a relatively inflexible machine can perform the work semi-autonomously.

Even when a task is fully codified, however, this does not mean it will be automated…. At the intersection of these two forces — technological feasibility and economic cost — lies the principle of comparative advantage."

95) Frey, C. B., & Osborne, M. A. (2013). The future of employment: How susceptible are jobs to computerisation. Retrieved on September 1, 2016 from http://www.oxfordmartin.ox.ac.uk/downloads/academic/The_Future_of_Employment.pdf

96) Carr, N. G. (2015). *The glass cage: How our computers are changing us*. New York, NY: W. W. Norton & Company.

97) Brody, R. G., Kowalczyk, T. K., & Coulter, J. M. (2003). The effect of a computerized decision aid on the development of knowledge. *Journal of Business and Psychology, 18*(2), 157-174.

98) McCall, H., Arnold, V., & Sutton, S. G. (2008). Use of knowledge management systems and the impact on the acquisition of explicit knowledge. *Journal of Information Systems, 22*(2), 77-101.

99) Sparrow, B., Liu, J., & Wegner, D. M. (2011). Google effects on memory: Cognitive consequences of having information at our fingertips. *Science, 333*(6043), 776-778.

100) Henkel, L. A. (2014). Point-and-shoot memories the influence of taking photos on memory for a museum tour. *Psychological science, 25*(2), 396-402.

101) Carr, N. G. (2015). *The glass cage: How our computers are changing us*. New York, NY: W. W. Norton & Company.

102) Brody, R. G., Kowalczyk, T. K., & Coulter, J. M. (2003). The effect of a computerized decision aid on the development of knowledge. *Journal of Business and Psychology, 18*(2), 157-174.

103) McCall, H., Arnold, V., & Sutton, S. G. (2008). Use of knowledge management systems and the impact on the acquisition of explicit knowledge. *Journal of Information Systems, 22*(2), 77-101.

104) van Nimwegen, C. (2008). *The paradox of the guided user: assistance can be counter-effective.* Utrecht, NL: Utrecht University.

105) van Nimwegen, C., & van Oostendorp, H. (2009). The questionable impact of an assisting interface on performance in transfer situations. *International Journal of Industrial Ergonomics, 39*(3), 501-508.

106) Burgos, D., & van Nimwegen, C. (2011) Games-based learning, destination feedback and adaptation: A case study of an educational planning simulation. In *Gaming and Simulations: Concepts, Methodologies, Tools and Applications* (pp. 1048-1059). Hershey, PA: IGI Global.

107) Dowling, C., Leech, S. A., & Moroney, R. (2008). Audit support system design and the declarative knowledge of long-term users. *Journal of Emerging Technologies in Accounting, 5*(1), 99-108.

108) Baxter, G., & Cartlidge, J. (2013). Flying by the seat of their pants: What can high frequency trading learn from aviation?. In *Proceedings of the 3rd International Conference on Application and Theory of Automation in Command and Control Systems* (pp. 64-73). Naples, Italy: ACM IRIT Press.

109) Haldar, V. (2014). Sharp tools, dull minds. Retrieved October 19, 2016, from http://blog.vivekhaldar.com/post/66660163006/sharp-tools-dull-minds

110) Burnett, G. E., & Lee, K. (2005). The effect of vehicle navigation systems on the formation of cognitive maps. In *International Conference of Traffic and Transport Psychology*.

111) Fenech, E. P., Drews, F. A., & Bakdash, J. Z. (2010, September). The effects of acoustic turn-by-turn navigation on wayfinding. In *Proceedings of the Human Factors and Ergonomics Society Annual Meeting*, 54(23), 1926-1930.

112) Maguire, E. A., Gadian, D. G., Johnsrude, I. S., Good, C. D., Ashburner, J., Frackowiak, R. S., & Frith, C. D. (2000). Navigation-related structural change in the hippocampi of taxi drivers. In *Proceedings of the National Academy of Sciences*, 97(8), 4398-4403.

# Chapter 5

1) Fleming, N. D. (1995, July). *I'm different; not dumb. Modes of presentation (VARK) in the tertiary classroom*. Paper presented at the 1995 Annual Conference of the Higher Education and Research Development Society of Australasia (HERDSA), 18, 308-313.

2) Fleming, N., & Baume, D. (2006). Learning styles again: VARKing up the right tree!. *Educational Developments*, 7(4), 4.

3) Bloom, B. S. (1956). *Taxonomy of educational objectives. Vol. 1: Cognitive Domain*. New York, NY: McKay.

4) Orey, M. (2010). Emerging perspectives on learning, teaching and technology. Retrieved October 20, 2016, from http://textbookequity. org/Textbooks/Orey_Emergin_Perspectives_Learning.pdf

5) McLoughlin, C., & Lee, M. J. (2010). Personalised and self-regulated learning in the web 2.0 era: International exemplars of innovative

pedagogy using social software. *Australasian Journal of Educational Technology*, 26(1), 28-43. Retrieved October 20, 2016, from http://www.todayonline.com/singapore/ challenge-maintaining-social-mobility-defies-easy-solutions-tharman

6) Sharples, M., Adams, A., Alozie, N., Ferguson, R., FitzGerald, E., Gaved, M., ... & Roschelle, J. (2015). Innovating pedagogy 2015: *Open university innovation report 4*. Milton Keynes: The Open University.

7) Roehl, A., Reddy, S. L., & Shannon, G. J. (2013). The flipped classroom: An opportunity to engage millennial students through active learning. *Journal of Family and Consumer Sciences, 105*(2), 44.

8) Gray, P. (2008, August 20). A brief history of education. In *Psychology Today*. Retrieved October 20, 2016, from https://www.psychologytoday. com/blog/freedom-learn/200808/brief-history-education

9) Goh, C. B., & Gopinathan, S. (2008). Education in Singapore: Development since 1965. In B. Fredriksen & J. P. Tan (Eds.), *An African Exploration of the East Asian Education* (pp. 80-108). Washington, DC: The World Bank.

10) World Economic Forum (2016, March). New vision for education: Fostering social and emotional learning through technology. Retrieved October 20, 2016, from http://www3.weforum.org/docs/ WEF_New_Vision_for_Education.pdf

11) Ministry of Education, Singapore (2015). 21st century competencies. Retrieved October 20, 2016, from https://www.moe.gov.sg/education/ education-system/21st-century-competencies

12) Ministry of Education, Singapore (n.d.). Values at the core of 21st century competencies. Retrieved October 20, 2016, from https://www.moe.gov.sg/docs/default-source/document/ education/21cc/files/annex-21cc-framework.pdf

13) Organisation for Economic Co-operation and Development. (n.d.). Global competency for an inclusive world. Retrieved October 20, 2016, from http://www.oecd.org/pisa/aboutpisa/ Global-competency-for-an-inclusive-world.pdf

14) Wagner, T., & Compton, R. A. (2015). *Creating innovators: The making of young people who will change the world.* New York, NY: Simon and Schuster.

15) Bootle, R. (2016, June 16). *Is the recovery just getting going – or running out of steam?* Speech presented at The Capital Economics Annual Conference, Singapore.

16) Williams, M. (2016, June 16). *Is the China crisis over?* Speech presented at The Capital Economics Annual Conference, Singapore.

17) Shah, S. (2016, June 16). *Is it time to re-think emerging markets?* Speech presented at The Capital Economics Annual Conference, Singapore.

18) Sharma, R. (2012). *Breakout nations: In pursuit of the next economic miracles.* New York, NY: W.W. Norton & Company.

19) Sharma, R. (2016). *The rise and fall of nations: Forces of change in the post-crisis world.* New York, NY: W. W. Norton & Company.

20) Cocco, F. (2016, August 18). Students opt for job-friendly A-levels. Retrieved October 20, 2016, from https://www.ft.com/content/ 6e8b7e9c-652a-11e6-8310-ecf0bddad227

21) Othman, L. (2016, May 26). Social mobility 'in trouble' as social gaps widen: Tharman. *Channel NewsAsia.* Retrieved October 20, 2016, from http://www.channelnewsasia.com/news/singapore/ social-mobility-in/2818920. html

22) Chin, N. C. (2016, May 27). Challenge of maintaining social mobility defies easy solutions: Tharman. *Today*. Retrieved October 20, 2016, from http://www.todayonline.com/singapore/ challenge-maintaining-social-mobility-defies-easy-solutions-tharman

23) Fletcher, J., & Wolfe, B. L. (2016). *The importance of family income in the formation and evolution of non-cognitive skills in childhood* (NBER Working Paper No. 22168). Cambridge, MA: National Bureau of Economic Research. doi: 10.3386/w22168

24) Gutman, L. M., & Schoon, I. (2013). The impact of non-cognitive skills on outcomes for young people. Education Endowment Foundation. Retrieved October 20, 2016, from http://educationendowmentfoundation.org.uk/uploads/pdf/ Non-cognitive_skills_literature_review.pdf

25) Vedanta, S. (2016, June 30). Researchers examine family income and children's non-cognitive skills. Retrieved October 20, 2016, from http://www.npr.org/2016/06/30/484129501/ researchers-examine-family-income-and-childrens-non-cognitive-skills

26) Asbury, K., & Plomin, R. (2013). *G is for genes: The impact of genetics on education and achievement*. New York, NY: John Wiley & Sons.

27) Adams, P. C. (1998). Teaching and learning with SimCity 2000. *Journal of Geography, 97*(2), 47-55.

28) Michael, D. R., & Chen, S. L. (2005). *Serious games: Games that educate, train, and inform*. Boston, MA: Thomson Course Technology.

29) Connolly, T. M., Boyle, E. A., MacArthur, E., Hainey, T., & Boyle, J. M. (2012). A systematic literature review of empirical evidence on computer games and serious games. *Computers & Education, 59*(2), 661-686.

30) The Economist (2016, June 15). The Economist explains: Predicting the success of "Hamilton". Retrieved October 20, 2016, from http://www.economist.com/blogs/economist-explains/2016/06/economist-explains-12

31) Flanagan, L. (2016, March 14). How Teachers Are Using 'Hamilton' the Musical in the Classroom. Retrieved January 15, 2017, from https://ww2.kqed.org/mindshift/2016/03/14/how-teachers-are-using-hamilton-the-musical-in-the-classroom/

32) Gorman. N. (2015, January 11) Broadway musical 'Hamilton' revitalizes civic education. Retrieved October 21, 2016, from http://www.educationworld.com/a_news/broadway-musical-'hamilton'-revitalizes-civic-education-157001260

33) Eastwood, J., & Hinton, E. (n.d.). How does 'Hamilton,' the non-stop, hip-hop broadway sensation tap rap's master rhymes to blur musical lines? Retrieved October 20, 2016, from http://graphics.wsj.com/hamilton/

34) Eastwood, J., & Hinton, E. (n.d.). How WSJ used an algorithm to analyze 'Hamilton' the musical. Retrieved October 20, 2016, from http://graphics.wsj.com/hamilton-methodology/

35) Carnegie Mellon University (n.d.) The CMU pronouncing dictionary. Retrieved October 20, 2016, from http://www.speech.cs.cmu.edu/cgi-bin/cmudict

36) Weisstein. E.W. (n.d.). Travelling salesman problem. Retrieved January 15, 2017, from http://mathworld.wolfram.com/TravelingSalesmanProblem.html

37) Kennedy, L. (2002, January 6). Spielberg in the twilight zone. Retrieved October 20, 2016, from https://www.wired.com/2002/06/spielberg/

38) Rodrigues, M., & Carvalho, P. S. (2013). Teaching physics with Angry Birds: exploring the kinematics and dynamics of the game. *Physics Education*, *48*(4), 431.

39) Mohanty, S. D., & Cantu, S. (2011). Teaching introductory undergraduate physics using commercial video games. *Physics Education*, *46*(5), 570.

40) Sun, C. T., Ye, S. H., & Wang, Y. J. (2015). Effects of commercial video games on cognitive elaboration of physical concepts. *Computers & Education*, *88*, 169-181.

41) Allan, R. (2010, August 10). The physics of Angry Birds. WIRED. Retrieved October 20, 2016, from https://www.wired.com/2010/10/physics-of-angry-birds/

42) McLeod, A., & Carabott, K. (2016, August 22). How Pokemon Go can be a positive tool for learning. *The Straits Times*. Retrieved October 20, 2016, from http://www.straitstimes.com/singapore/education/how-pokemon-go-can-be-a-positive-tool-for-learning

43) Modafferi. M. (2016, July 18). 5 ways to trick students into learning with Pokemón Go. Retrieved October 20, 2016, from https://blog.education.nationalgeographic.com/2016/07/18/4-ways-to-trick-students-into-learning-with-pokemon-go/

44) Gorman, N. (2016, July 14). Education world: How Pokémon GO is the perfect tool for encouraging summer learning. Retrieved October 20, 2016, from http://www.educationworld.com/a_news/how-pokemon-go-perfect-tool-encouraging-summer-learning-1687680210

45) LINKX-app. (n.d.). Retrieved October 20, 2016, from http://www.facebook.com/LINKXapp

46) Wilson, M. (2016, December 12). "Pokémon Go" is quietly helping people fall in love with their cities. Retrieved October 20, 2016, from https://www.fastcodesign.com/3061718/pokemon-go-is-quietly-helping-people-discover-their-cities

47) Poon, K. W., Tay, E., Chae, Y., Balasubramanian, G., & Yong, A. (2016). Teaching economics in 2020 - What it means for 2040. *Economics & Society, Man and Environment, 1*, 26-31.

48) Gardner, H. (2006). *Five minds for the future*. Cambridge, MA: Harvard Business Press.

49) Levina, N., & Vaast, E. (2005). The emergence of boundary spanning competence in practice: implications for implementation and use of information systems. *MIS Quarterly, 29*(2), 335-363.

50) Azoulay, P., Zivin, J. G., & Manso, G. (2011). Incentives and creativity: Evidence from the Howard Hughes medical investigator program. *The RAND Journal of Economics, 42*(3), 527-554.

51) Lee, H. (2016, August 12). EduBang video explanation - revised [Video file]. Retrieved October 22, 2016, from https://www.youtube.com/watch?v=OSl7utAYbOU

52) Susskind, R. E., & Susskind, D. (2016). *The future of the professions: How technology will transform the work of human experts*. New York, NY: Oxford University Press.

The authors discuss "[e]merging skills and competencies" such as different ways of communication with new technologies, mastery of data, how to work alongside technology, to know when technology can do a better job, and to diversify and extend their expertise into new disciplines (often with the help of technologies). These are effectively new "digital literacies".

53) Tapscott, D. (2009). *Grown up digital*. New York, NY: McGraw-Hill.

54) Pollack, L. (2016, August 17). Ageing out of the 25-34 bracket, one app at a time. Retrieved October 20, 2016, from https://www.ft.com/content/de0e26c8-62f2-11e6-a08a-c7ac04ef00aa#axzz4IngM2TqJ

55) Topping, K. J., Dehkinet, R., Blanch, S., Corcelles, M., & Duran, D. (2013). Paradoxical effects of feedback in international online reciprocal peer tutoring. *Computers & Education*, *61*, 225-231.

56) Tsuei, M. (2012). Using synchronous peer tutoring system to promote elementary students' learning in mathematics. *Computers & Education*, *58*(4), 1171-1182.

57) Evans, M. J., & Moore, J. S. (2013). Peer tutoring with the aid of the Internet. *British Journal of Educational Technology*, *44*(1), 144-155.

58) Tolosa, C., East, M., & Villers, H. (2016). Online peer feedback in beginners' writing tasks: Lessons learned. *IALLT Journal of Language Learning Technologies*, *43*(1), 1-24.

59) Yang, S., & Wahab, N. (2016, March 9). Reading without borders: Volunteer in Singapore helps student in Malaysia learn English. *Channel NewsAsia*. Retrieved October 20, 2016, from http://www.channelnewsasia.com/news/singapore/reading-without-borders/2579892.html

As the technologies used improve (e.g. feedback, communications AI, etc.), we expect the potential benefits to expand too.

60) Yenn, T. Y. (2016, March 10). Why low-income parents may make 'poor choices'. *The Straits Times*. Retrieved October 20, 2016, from http://www.straitstimes.com/opinion/why-low-income-parents-may-make-poor-choices

61) Ministry of Education, Singapore. (n.d.). ICT Masterplan 4. Retrieved October 20, 2016, from http://ictconnection.moe.edu.sg/masterplan-4/overview

62) Examples of ICT used in schools (from our interviews and public sources)

63) Stanford University (2016, September). "Artificial Intelligence and Life in 2030." One hundred year study on artificial intelligence: Report of the 2015-2016 study panel. Stanford, CA: Stanford University. Retrieved October 20, 2016, from https://ai100.stanford.edu/2016-report

64) The Economist. (2016, June 9). Teaching the teachers. Retrieved October 20, 2016, from http://www.economist.com/news/briefing/21700385-great-teaching-has-long-been-seen-innate-skill-reformers-are-showing-best

65) Singer, N. (2015, September 5). A sharing economy where teachers win. Retrieved October 20, 2016, from http://www.nytimes.com/2015/09/06/technology/a-sharing-economy-where-teachers-win. html

66) Bloom, B. S. (1984). The 2 sigma problem: The search for methods of group instruction as effective as one-to-one tutoring. *Educational researcher, 13*(6), 4-16.

67) Aloisi, A. (2015). Commoditized workers the rising of on-demand work, a case study research on a set of online platforms and apps. Retrieved October 20, 2016, from http://ssrn.com/abstract=2637485

68) The Economist (2015, January 03). The on-demand economy - Workers on tap. Retrieved October 19, 2016, from http://www.economist.com/news/leaders/21637393-rise-demand-economy-poses-difficult-questions-workers-companies-and

69) O'Connor, S. (2015, October 9). The human cloud: A new world of work. Retrieved October 19, 2016, from http://www.ft.com/cms/s/2/a4b6e13e-675e-11e5-97d0-1456a776a4f5.html#axzz4J0jv232W

70) Jagdish, H. (2016, February 27). Singaporean musicians need to find ways to be Singaporean. *Channel NewsAsia*. Retrieved October 20, 2016, from http://www.channelnewsasia.com/news/singapore/singaporean-musicians/2552856.html

71) Teng, A. (2014, October 23). Singapore maths is travelling the world. *The Straits Times*.

72) Pivotplanet (n.d.). Retrieved October 20, 2016, from https://www.pivotplanet.com/

73) Institute for Personal Leadership. (n.d.). Virtual mentor (Pilot). Retrieved October 20, 2016, from http://www.personalleadership.com/virtual-mentor-pilot

74) Ilyer, B. & Murphy, W. (2016, April 26). The Benefits of Virtual Mentors. *Harvard Business Review*. Retrieved October 20, 2016, from https://hbr.org/2016/04/the-benefits-of-virtual-mentors

75) [Personal interview]. (2016).
During an interview, the interviewers and interviewees joked that bringing the professions into school education would be akin to having a KidZania for school.

76) Sin, Y. (2016, July 11). 'Externships' growing in popularity in Singapore. The Straits Times. Retrieved October 20, 2016, from http://www.straitstimes.com/singapore/education/externships-growing-in-popularity-here

At one level, one could argue this is taking vocationalisation of education to its extremes – students are already embarking on multiple internships (one student we interviewed said his advice to his juniors was "do internships at every opportunity to explore your options") and externships, and schools at different levels are also encouraging industry exposure programs.

77) Poon, K. W., Tay, E., Chae, Y., Balasubramanian, G., & Yong, A. (2016). Teaching economics in 2020 - What it means for 2040. *Economics & Society, Man and Environment, 1*, 26-31.

78) Lee, H. (2016, August 12). EduBang video explanation - revised [Video file]. Retrieved October 22, 2016, from https://youtu.be/8QkqNnKlGPA

# Chapter 6

1) Ministry of Health, Singapore. (2016, August 22). Singapore health facts. Retrieved October 23, 2016, from http://www.moh.gov.sg/content/moh_web/home/statistics/Health_Facts_Singapore.html

2) Ministry of Health, Singapore. (2015, October 15). Hospital services. Retrieved October 23, 2016, from http://www.moh.gov.sg/content/moh_web/home/our_healthcare_system/Healthcare_Services/Hospitals.html

3) Ministry of Health, Singapore. (2016). Singapore health facts. Retrieved October 24, 2016, from https://www.moh.gov.sg/content/moh_web/home/statistics/Health_Facts_Singapore/Health_Facilities.html

4) Ministry of Education, Singapore. (2015). *Education Statistics Digest 2015*. Retrieved October 23, 2016, from

https://www.moe.gov.sg/docs/default-source/document/publications/
education-statistics-digest/esd-2015.pdf

5) Cai, H. X. (2014, March 24). A stroll through the shopping mall sector.
*The Business Times*. Retrieved October 24, 2016, from
http://www.btinvest.com.sg/personal_finance/young-investors-forum/
a-stroll-through-the-shopping-mall-sector/

There are more than 80 malls in Singapore, scattered over all parts of
the island, usually located near an MRT station.

6) Department of Statistics, Singapore. (n.d.). Retrieved October 23,
2016, from http://www.singstat.gov.sg/statistics/

7) Department of Statistics, Singapore. (2016, October 17). Statistics
Singapore - Latest data. Retrieved October 23, 2016, from
http://www.singstat.gov.sg/statistics/latest-data

8) Department of Statistics, Singapore. (2015). Profile of enterprises in
Singapore. Retrieved October 23, 2016, from
https://www.singstat.gov.sg/docs/default-source/default-document-
library/statistics/visualising_data/profile-of-enterprises-2015.pdf

9) Department of Statistics, Singapore. (n.d.). Retrieved October 23,
2016, from http://www.singstat.gov.sg/

10) Giraldo, J.P., Landry, M.P., Faltermeier, S.M., McNicholas, T.P.,
Iverson, N.M., Boghossian, A.A., ... & Strano, M.S. (2014). Plant
nanobionics approach to augment photosynthesis and biochemical
sensing. *Nature Materials, 13*(4), 400-408.

11) Trafton, A. (2014, March 16). Bionic plants. *MIT News*. Retrieved
October 20, 2016, from http://news.mit.edu/2014/bionic-plants

12) Bhavnani, S. P., Narula, J., & Sengupta, P. P. (2016). Mobile technology and the digitization of healthcare. *European Heart Journal*, 37(18), 1428-1438.

13) Misfit & Digital Health group on LinkedIn (2014, August) health infographic.jpg [Online image]. (n.d.). Retrieved October 21, 2016, from http://innotechtoday.com/wp-content/uploads/2014/08/graph.png

14) The Economist. (2012, December 1). The dream of the medical tricorder. Retrieved October 21, 2016, from http://www.economist.com/news/technology-quarterly/21567208-medical-technology-hand-held-diagnostic-devices-seen-star-trek-are-inspiring

15) Qualcomm tricorder XPRIZE. (n.d.). Retrieved October 21, 2016, from http://tricorder.xprize.org/

16) Cyrcadia Health (n.d.) Cyrcadia health-early detection technology for breast cancer. Retrieved October 20, 2016, from http://cyrcadiahealth.com/

"The iTBra consists of two wearable, comfortable intelligent breast patches which detect circadian temperature changes within breast tissue. Through your smartphone or PC, anonymized data obtained from the iTBra is communicated directly to the Cyrcadia Health core lab for analysis. Developed in conjunction with the world class Nanyang Technological University of Singapore, the Cyrcadia Health solution employs machine learning predictive analytic software, a series of algorithms to identify and categorize abnormal circadian patterns in otherwise healthy breast tissue. Once the data is submitted, Cyrcadia Health will deliver accurate, reproducible and automated results to health care providers automatically and within minutes."

17) Langley, L. (2016, March 19). How dogs can sniff out diabetes and cancer. *National Geographic*. Retrieved October 20, 2016, from

http://news.nationalgeographic.com/2016/03/
160319-dogs-diabetes-health-cancer-animals-science/

"Numerous studies have shown man's best friend can detect various cancers, including prostate cancer, colorectal cancer and melanoma."

18) Taverna, G., Tidu, L., Grizzi, F., Torri, V., Mandressi, A., Sardella, ... & Graziotti, P. (2015). Olfactory system of highly trained dogs detects prostate cancer in urine samples. *The Journal of Urology, 193*(4), 1382-1387.

19) Sonoda, H., Kohnoe, S., Yamazato, T., Satoh, Y., Morizono, G., Shikata, K., ... & Inoue, F. (2011). Colorectal cancer screening with odour material by canine scent detection. *Gut, 60*, 814-819. doi:10.1136/gut.2010.218305

20) Pickel, D., Manucy, G. P., Walker, D. B., Hall, S. B., & Walker, J. C. (2004). Evidence for canine olfactory detection of melanoma. *Applied Animal Behaviour Science, 89*(1), 107-116.

21) Scutti, S. (2016, July 26). Does it pass the 'smell test'? Seeking ways to diagnose Alzheimer's early. Retrieved October 20, 2016, from http://edition.cnn.com/2016/07/26/health/alzheimers-eye-and-smell/

"Two studies presented Tuesday at the Alzheimer's Association International Conference 2016 suggested that older adults with worsening ability to identify odors might be on the road to cognitive decline. Two other presentations explored different types of eye tests as possible predictors of the disease."

22) Solon, O. (2014, December 9). Does your child have eye cancer? Use a smartphone. Retrieved October 20, 2016, from http://www.mirror.co.uk/news/technology-science/technology/your-child-eye-cancer-use-4774686

23) Alzheimer's Society. (2016, July 26). One step closer to eye tests to reveal early signs of memory problems and dementia. Retrieved October 21, 2016, from https://www.alzheimers.org.uk/news/article/85/ one_step_closer_to_eye_tests_to_reveal_early_signs_of_memory_ problems_and_dementia

24) Vincent, J. (2016, July 5). Google DeepMind will use machine learning to spot eye diseases early. Retrieved October 20, 2016, from http://www.theverge.com/2016/7/5/12095830/ google-deepmind-nhs-eye-disease-detection

25) Stanford Medicine. (2014, November 5). Retinal-scan analysis can predict advance of macular degeneration, study finds. Retrieved October 20, 2016, from https://med.stanford.edu/news/all-news/2014/11/ retinal-scan-analysis-can-predict-advance-of-macular-degeneratio.html

26) Haiken, M. (2014, July 16). Can an eye test predict Alzheimer's? Scientists unveil new vision scans. Retrieved October 20, 2016, from http://www.forbes.com/sites/melaniehaiken/2014/07/16/ a-simple-vision-test-for-alzheimers-scientists-unveil-new-technologies/ #3c2968b942b1

27) Singapore National Eye Centre (2010, December 11) Doc pioneers eye scan that can predict disease. Retrieved October 20, 2016, from http://www.snec.com.sg/about/newsroom/news-articles/Pages/ 11Dec2010_TheStraitsTimes_PgD2.aspx

28) Ventola, C. L. (2014). Mobile devices and apps for health care professionals: uses and benefits. *Pharmacy and Therapeutics*, *39*(5), 356.

29) Laksanasopin, T., Guo, T.W., Nayak, S., Sridhara, A.A., Xie, S., Olowookere, O.O., ... & Chin, C.D. (2015). A smartphone dongle for diagnosis of infectious diseases at the point of care. *Science Translational Medicine*, *7*(273), 273re1.

30) Desmon, S. (February 13). 'Mini-brains' developed at Johns Hopkins could reshape brain research, drug testing. Retrieved October 20, 2016, from http://hub.jhu.edu/2016/02/12/mini-brains-drug-testing/

31) Ledford, H. (2016, March 7). CRISPR: Gene editing is just the beginning : Nature news ... Retrieved October 20, 2016, from http://www.nature.com/news/ crispr-gene-editing-is-just-the-beginning-1.19510

32) Donati, A.R., Shokur, S., Morya, E., Campos, D.S., Moioli, R.C., Gitti, C.M., ... & Brasil, F.L. (2016). Long-term training with a brain-machine interface-based gait protocol induces partial neurological recovery in paraplegic patients. *Scientific Reports*, 6.

33) Heater, B. (2016, August 12). Using VR and an exoskeletons to help paraplegics regain movement. Retrieved October 20, 2016, from https://techcrunch.com/2016/08/12/duke-study/

34) Khalil, S. (2016, August 24). More in Singapore surviving cancer battle. *The Straits Times*. Retrieved October 21, 2016, from http://www.straitstimes.com/singapore/more-surviving-cancer-battle

35) Baker, D.J., Childs, B.G., Durik, M., Wijers, M.E., Sieben, C.J., Zhong, J., ... & Khazaie, K. (2016). Naturally occurring p16Ink4a-positive cells shorten healthy lifespan. *Nature*, *530*(7589), 184-189.

36) Mayo Clinic. (2016, February 3). Lifespan of mice extended by as much as 35 percent; no adverse effects found. Retrieved October 21, 2016, from https://www.sciencedaily.com/releases/2016/02/ 160203145723.htm

37) Sandel, M. J. (2009). *The case against perfection*. Cambridge, MA: Harvard University Press.

38) Gille, L., & Houy, T. (2014). The future of health care demand in developed countries: From the "right to treatment" to the "duty to stay healthy". *Futures, 61*, 23-32.

39) Kraft, D. (2016, August 22). The future of healthcare is arriving — 8 exciting areas to watch. Retrieved October 21, 2016, from http://singularityhub.com/2016/08/22/exponential-medicine-2016-the-future-of-health-care-is-coming-faster-than-you-think/

40) Arivale. (n.d.). Arivale - Your scientific path to wellness. Retrieved October 21, 2016, from https://www.arivale.com/

41) Human Longevity, Inc. (n.d.). Products and services. Retrieved October 21, 2016, from http://www.humanlongevity.com/science-technology/products/

42) Vincent, J. (2015, May 4). Smartphones can detect eye cancer. Retrieved October 21, 2016, from http://www.theverge.com/2015/5/14/8604923/eye-cancer-smartphone-diagnosis-white-glow

43) DeMiranda, M.A., Doggett, A.M. & Evans, J.T. (2005). Medical technology: Context and content in science and technology. Retrieved on October 26, 2016 from https://www.researchgate.net/profile/Mark_Doggett/publication/42831446_Medical_Technology_Contexts_and_Content_in_Science_and_Technology/links/02e7e5367ee83d510a000000.pdf

44) Islam, S. R., Kwak, D., Kabir, M. H., Hossain, M., & Kwak, K. S. (2015). The internet of things for health care: a comprehensive survey. *IEEE Access, 3*, 678-708.

45) Hibbard, J., & Gilburt, H. (2014). Supporting people to manage their health: an introduction to patient activation. London: The King's Fund.

Retrieved October 26, 2016 from http://www.kingsfund.org.uk/
publications/supporting-people-manage-their-health

46) World Health Organization. (2005). *Preventing chronic diseases:
a vital investment*. Geneva: World Health Organization.

47) Chandler, A. D. (2009). *Shaping the industrial century: The
remarkable story of the evolution of the modern chemical and
pharmaceutical industries*. Cambridge, MA: Harvard University Press.

48) Chandler, A. D., Hikino, T., & Von Nordenflycht, A. (2005). *Inventing
the electronic century: The epic story of the consumer electronics and
computer industries, with a new preface*. Cambridge, MA: Harvard
University Press.

49) Brookings Institution (2016, July 19). The 5G network, the internet
of things, and the future of health care: Opening remarks [Video
file]. Retrieved October 21, 2016, from https://www.youtube.com/
watch?v=OFLayTwVXlw

50) War on cancer - Wikipedia. (n.d.). Retrieved October 21, 2016, from
https://en.wikipedia.org/wiki/War_on_Cancer

51) Cities Changing Diabetes. (n.d.). Retrieved October 21, 2016, from
http://citieschangingdiabetes.com/

52) Khalil, S. (2016, April 23). Parliament: Health minister Gan Kim Yong
declares 'war on diabetes'; new task force set up. *The Straits Times*. Retrieved October 21, 2016, from http://www.straitstimes.com/singapore/
health/moh-declares-war-against-diabetes

53) Nesta. (n.d.). What we have learnt from People Powered Health.
Retrieved October 21, 2016, from http://www.nesta.org.uk/
what-we-have-learnt-people-powered-health

54) Ibid.

55) Bland, J., Khan, H., Loser, J., Simons, T., & Westlake, S. (2015, July). The NHS in 2030. Retrieved October 21, 2016, from https://www.nesta.org.uk/sites/default/files/the-nhs-in-2030.pdf

56) Hibbard, J., & Gilburt, H. (2014). Supporting people to manage their health: an introduction to patient activation. London: The King's Fund. Retrieved October 26, 2016 from http://www.kingsfund.org.uk/ publications/supporting-people-manage-their-health

"'Patient activation' … describes the knowledge, skills and confidence a person has in managing their own health and health care… Patients with low activation levels are more likely to attend accident and emergency departments, to be hospitalised or to be re-admitted to hospital after being discharged… Patient activation is a powerful mechanism for tackling health inequalities. Used in population segmentation and risk stratification, it provides new insights into risk that go beyond those obtained using traditional socio-demographic factors. Patient activation provides a unique measure of engagement and empowerment hat can be used to evaluate the effectiveness of interventions and to measure the performance of health care organisations in involving patients in their own care."

57) Greene, J., & Hibbard, J. H. (2012). Why does patient activation matter? An examination of the relationships between patient activation and health-related outcomes. *Journal of General Internal Medicine, 27*(5), 520-526.

58) National Healthcare Group Polyclinics. (2013). Transform Care. Retrieved October 21, 2016, from https://www.nhgp.com.sg/WorkArea/ DownloadAsset.aspx?id=2557

Patient Activation. See also "Dr Darren Seah - Photograph-Assisted Dietary Review amongst Type 2 Diabetics in Primary Care" where "A research study at NHGP [in Singapore] shows that photographic records of patients' dietary intake help patients better manage their condition....The Patient Activation Measure (PAM) scores – a scale to gauge the knowledge, skills and confidence essential to managing one's own health – have also increased significantly. This indicated that patients demonstrated better self-management of the condition."

59) Taltioni. (n.d.). Sitra. Retrieved October 21, 2016, from http://www.sitra.fi/en/taltioni

"[Finland's] national Taltioni service, which provides a single platform for the storage of information on the health and well-being of Finns. It can be used by health care providers - in either the public or private sector - and covers all aspects of health; from the treatment of illnesses, to the promotion of general well-being and the spread of advice on how to prevent ill health.

With Taltioni, people get tangible personal tools for the maintenance of their own health and well-being.
- Taltioni lets people save, collect, produce, use and share information on their own health and well-being
- Health and well-being information can be shared and used independent of time and location
- The health data is safe and secure
- Information on the person remains in his or her own control despite changing jobs, finishing studies or moving to a new town

You own your information and can share selected information with other users or health care professionals as you see fit. For example, shared information can be used to monitor the well-being of a family member. Examples where the Taltioni service can be of use include the treatment of chronic diseases, health and exercise journals, health coaching and delivery of laboratory results."

60) SitraFund. (2010, December 26). Taltioni - Finnish PHR platform [Video file]. Retrieved October 21, 2016, from https://www.youtube.com/watch?v=5h-W7PP9K1o

61) Topol, E. J. (2015). *The patient will see you now: The future of medicine is in your hands*. New York, NY: Basic Books.

62) Haseltine, W. A. (2013). *Affordable excellence: the Singapore healthcare story*. Washington, DC: Brookings Institution Press.

63) Lee, C. E., & Satku, K. (Eds.). (2015). *Singapore's health care system: What 50 years have achieved.* Singapore: World Scientific.

64) Simon, H. A. (1956). Rational choice and the structure of the environment. *Psychological Review, 63*(2), 129.

65) Agency for Healthcare Research and Quality. (n.d.). 5. Improving data collection across the health care system. Race, ethnicity, and language data: Standardization for health care quality improvement. Retrieved October 21, 2016, from http://www.ahrq.gov/research/findings/final-reports/iomracereport/reldata5.html

66) Nature (2013, December 5). The FDA and me. *Nature, 504*, 7-8. doi: 10.1038/504007b

67) Topol, E. J. (2015). *The patient will see you now: The future of medicine is in your hands*. New York, NY: Basic Books.

From p. 72:
"Creating a massive information resource of millions of customers, and monetizing the anonymized data for pharmaceutical companies or

health insurers, represents an unproven strategy, not just because they need customers to buy the data, but because they must protect the data to avoid re-identification… some people are going to be very uncomfortable with the prospect of selling their information to a drug company."

68) PatientsLikeMe. (2009, June 15). UCB and PatientsLikeMe partner to give people with Epilepsy a voice in advancing research. Retrieved January 15, 2017, http://www.marketwired.com/press-release/ucb-patientslikeme-partner-give-people-with-epilepsy-voice-advancing-research-1221273.htm

PatientsLikeMe is an online patient network headquartered in Cambridge, Massachusetts. Its website was launched on October 10, 2005 with the goal of connecting patients with one another, improving their outcomes, and enabling research.

69) ClinGen. (n.d.). Sharing Clinical Reports Project (SCRP). Retrieved October 21, 2016, from https://www.clinicalgenome.org/data-sharing/sharing-clinical-reports-project-scrp/

Sharing Clinical Reports Project (SCRP) is a volunteer, grass-roots effort to encourage open sharing of breast cancer, BRCA 1|2 gene variant information.

70) Ayres, I. (2007). *Super crunchers: Why thinking-by-numbers is the new way to be smart.* New York, NY: Bantam Books.

71) Mayo Clinic. (n.d.). *Sniffers - Clinical informatics in intensive care lab.* Retrieved October 21, 2016, from http://www.mayo.edu/research/labs/clinical-informatics-intensive-care/projects/sniffers

"Using near-real-time data feeds from Mayo Clinic's electronic medical record and the intensive care unit (METRIC) datamart, the laboratory of Vitaly Herasevich, M.D., Ph.D., and Brian W. Pickering, M.B., B.Ch., has developed and tested a number of highly sensitive and specific electronic alerts (sniffers). The laboratory has developed and validated sniffers for acute lung injury, ventilator-induced lung injury and septic shock. Sniffers for ventilator-associated pneumonia, acute kidney injury and a number of other conditions are under development."

72) World Health Organisation. (n.d.). Out-of-pocket expenditure on health as a percentage total expenditure on health (US$) (n.d.). Retrieved October 21, 2016, from http://www.who.int/gho/ health_financing/out_pocket_expenditure_total/en/

73) Armerding, T. (2016, April 22). Uber fraud: Scammer takes the ride, victim gets the bill. Retrieved October 21, 2016, from http://www.csoonline.com/article/3059461/data-breach/ uber-fraud-scammer-takes-the-ride-victim-gets-the-bill.html

74) Pentland, A. (2013). How big data can transform society for the better. *Scientific American*, *309*(4).

Pentland writes that our daily "digital traces... could become a privacy nightmare - or it could be the foundation of a healthier, more prosperous world."

75) Delbanco, T., Walker, J., Bell, S.K., Darer, J.D., Elmore, J.G., Farag, N., ... & Ross, S.E. (2012). Inviting patients to read their doctors' notes: a quasi-experimental study and a look ahead. *Annals of internal medicine*, *157*(7), 461-470.

76) Simon, H. A. (1956). Rational choice and the structure of the environment. *Psychological Review*, *63*(2), 129.

77) Simonite, T. (2014, September 8). Datacoup wants to buy your credit card and Facebook data. Retrieved October 24, 2016, from

https://www.technologyreview.com/s/530486/
datacoup-wants-to-buy-your-credit-card-and-facebook-data/

"Datacoup will pay up to $10 for access to your social network accounts, credit card transaction records, and other personal information, and will sell insights gleaned from that data to companies looking for information on consumer behavior… Whether an individual user gets the full $10 a month or not depends on which streams of data he's willing to share. Options include debit card and credit card transactions, and data from Facebook, Twitter, and LinkedIn."

78) Ross, W. (2014, August 26). How much is your privacy worth? *MIT Technology Review*. Retrieved October 24, 2016, from https://www. technologyreview.com/s/529686/how-much-is-your-privacy-worth/

"Despite the outcry over government and corporate snooping, some people allow themselves to be monitored for money or rewards."

79) Our data mutual. (n.d.). Retrieved October 24, 2016, from http://www.ourdatamutual.org/#how-it-works

80) Healthbank. (2016). Retrieved October 24, 2016, from http://www.healthbank.coop/

The introduction goes: "We empower people across the globe to exchange their health data on our uniquely neutral and independent platform. Healthbank drives innovation in health sciences, from prevention to cure, at a better price with better quality for the benefit of both the individual and society."

81) Tanner, A. (2016, February 1). How data brokers make money off your medical records. Retrieved October 24, 2016, from https://www.scientificamerican.com/article/
how-data-brokers-make-money-off-your-medical-records/

82) Torres, C. D. (2014, February 19). You want my data? What's it worth? Retrieved October 24, 2016, from http://www.huffingtonpost.ca/ carlos-de-torres-gimeno/big-data_b_4817705.html

83) Steel, E., Locke, C., Cadman, E., & Freese, B. (2013, June 12). How much is your personal data worth? Retrieved October 24, 2016 from http://www.ft.com/cms/s/2/927ca86e-d29b-11e2-88ed-00144feab7de.html?ft_site=falcon#axzz2wSAnBCdm

FT created an online calculator for estimating how much your data is worth.

84) Pettypiece, S., & Robertson, J. (2014, September 19). For sale: Your name and medical condition. Bloomberg. Retrieved October 24, 2016, from http://www.bloomberg.com/news/articles/2014-09-18/ for-sale-your-name-and-medical-condition

85) theDataMap. (2016). Retrieved October 24, 2016, from http://thedatamap.org/map2013/index.php

86) theDataMap. (2016). Others can learn your medical details. Retrieved October 24, 2016, from http://thedatamap.org/ map2013/staterisks.php

The researchers also claimed that they paid $50 for a public dataset which was anonymized and showed how it's possible to still identify individuals in it. Such datasets are available for sale to companies.

87) Patient Privacy Rights. (2012). Personal health data is for sale - Patient privacy rights. Retrieved October 24, 2016, from http://patientprivacyrights.org/wp-content/uploads/ 2012/10/PPR-Data-for-Sale-Peel.pdf

88) McGowan, K. (2016, February 11). Our medical data must become free. Retrieved October 24, 2016, from https://backchannel.com/our-medical-data-must-become-free-f6d533db6bed#.cjbn8rirq

"Data generated by your body is routinely captured and sold by healthcare companies. Shouldn't you benefit from it, too? While a stolen credit card account is worth just a few bucks, medical records can sell for $10 to $350 each, depending on the security expert you ask."

89) Humer, C., & Finkle, J. (2014, September 24). Your medical record is worth more to hackers. Retrieved October 24, 2016, from http://www.reuters.com/article/us-cybersecurity-hospitals-idUSKCN-0HJ21I20140924

"Stolen health credentials can go for $10 each, about 10 or 20 times the value of a U.S. credit card number, according to Don Jackson, director of threat intelligence at PhishLabs, a cybercrime protection company. He obtained the data by monitoring underground exchanges where hackers sell the information."

90) Mangan, D. (2016, March 9). As health data breaches increase, what do you have to lose? CNBC. Retrieved October 24, 2016, from http://www.cnbc.com/2016/03/09/as-health-data-breaches-increase-what-do-you-have-to-lose.html

"...a single medical record tied to an individual can now sell for "up to $1,100.".....About two years ago, it was probably worth no more than $50."

91) Rashid, F. Y. (2015, September 14). Why hackers want your health care data most of all. Retrieved October 24, 2016, from http://www.infoworld.com/article/2983634/security/why-hackers-want-your-health-care-data-breaches-most-of-all.html

"The FBI said recently criminals can sell health care information for as much as $50 a record."

92) Walker, J., Leveille, S.G., Ngo, L., Vodicka, E., Darer, J.D., Dhanireddy, S., ... & Ralston, J.D. (2011). Inviting patients to read their doctors' notes: patients and doctors look ahead: patient and physician surveys. *Annals of Internal Medicine*, *155*(12), 811-819.

93) Topol, E. J. (2015). *The patient will see you now: The future of medicine is in your hands*. New York, NY: Basic Books.

From p. 135:
"Patient-generated data will, in the years ahead, become the largest and most diverse of all. It encompasses wearable biosensors, imaging, and laboratory tests, including genomics and other omics."

From p. 195:
"When you have gigabytes of data, perhaps hundreds of gigabytes, for each patient, that's more data than has existed in all clinical trials combined up until a couple of years ago."

From p. 246:
"Currently the annual amount of data produced worldwide per individual is about one terabyte... Just the omics from an individual will add at least another five terabytes, and we haven't even gotten to real-time streaming from biosensors..."

94) Smith, M., Saunders, R., Stuckhardt, L., & McGinnis, J. M. (Eds.). (2013). *Best care at lower cost: the path to continuously learning health care in America*. Washington, DC: National Academies Press.

95) Katzan, I. L., & Rudick, R. A. (2012). Time to integrate clinical and research informatics. *Science Translational Medicine*, *4*(162), 162fs41-162fs41.

96) Kahn, J. S., Aulakh, V., & Bosworth, A. (2009). What it takes: characteristics of the ideal personal health record. *Health Affairs, 28*(2), 369-376.

97) Kotz, J. (2012). Bringing patient data into the open. *SciBX: Science-Business eXchange, 5*(25).

98) Kierkegaard, P. (2013). eHealth in Denmark: a case study. *Journal of Medical Systems, 37*(6), 1-10.

99) Nature (n.d.). Careless data. *Nature*. Retrieved October 21, 2016, from http://www.nature.com/news/careless-data-1.14802

100) Competition and Markets Authority, UK. (2015, February 26). Online reviews and endorsements. Retrieved October 21, 2016, from https://www.gov.uk/cma-cases/online-reviews-and-endorsements

101) U.S. National Library of Medicine. (n.d.). What is the difference between precision medicine and personalised medicine? What about pharmacogenomics? Retrieved October 21, 2016, from https://ghr.nlm.nih.gov/primer/precisionmedicine/precisionvspersonalized

102) Hibbard, J., & Gilburt, H. (2014). Supporting people to manage their health: an introduction to patient activation. London: The King's Fund. Retrieved October 26, 2016 from http://www.kingsfund.org.uk/publications/supporting-people-manage-their-health

103) Nijman, J., Hendriks, M., Brabers, A., de Jong, J., & Rademakers, J. (2014). Patient activation and health literacy as predictors of health information use in a general sample of Dutch health care consumers. *Journal of Health Communication, 19*(8), 955-969.

104) Hibbard, J., & Gilburt, H. (2014). Supporting people to manage their health: an introduction to patient activation. London: The King's Fund.

Retrieved October 26, 2016 from http://www.kingsfund.org.uk/ publications/supporting-people-manage-their-health

105) Ibid.

106) Thaler, R. H., & Sunstein, C. R. (2008). *Nudge: Improving decisions about health, wealth, and happiness.* New Haven, CT: Yale University Press.

107) Ministry of Health, Singapore. (n.d.). National Electronic Health Record (NEHR). Retrieved October 21, 2016, from https://www.moh.gov.sg/content/moh_web/home/Publications/ educational_resources/2015/national-electronic-health-record--nehr-.html

108) APAC CIO Outlook (2015, August). Holmusk: HealthCare solutions by leveraging big data. Retrieved October 21, 2016, from http://bigdata.apacciooutlook.com/vendor/article5/holmusk

109) Barr, A. (2014, July 27). Google's new moonshot project: The human body. *The Wall Street Journal.* Retrieved October 21, 2016, from http://www.wsj.com/articles/google-to-collect-data-to-define-healthy-human-1406246214

110) Richter, R. (2014, July 25). Stanford partnering with Google [x] and Duke to better understand the human body. Retrieved October 21, 2016, from http://scopeblog.stanford.edu/2014/07/25/stanford-partnering-with-google-x-and-duke-to-better-understand-the-human-body/

111) Advanced Functional Fabrics of America. (n.d.). About – AFFOA. Retrieved October 21, 2016, from http://join.affoa.org/about/

112) Dacanay, M. (2016, April 1). Public private consortium pours $317 million for advanced functional fibers of America: What the project is about. Retrieved October 21, 2016, from http://www.techtimes.com/

articles/146287/20160401/public-private-consortium-pours-
317-million-advanced-functional-fibers-america.htm

113) Islam, S. R., Kwak, D., Kabir, M. H., Hossain, M., & Kwak, K. S.
(2015). The internet of things for health care: a comprehensive survey.
*IEEE Access*, *3*, 678-708.

114) Pentland, A. (2014). *Social physics: How good ideas spread - The
lessons from a new science*. New York, NY: Penguin.

115) Pickard, G., Pan, W., Rahwan, I., Cebrian, M., Crane, R., Madan,
A., & Pentland, A. (2011). Time-critical social mobilization.
*Science*, *334*(6055), 509-512.

116) Mani, A., Rahwan, I., & Pentland, A. (2013). Inducing peer pressure
to promote cooperation. *Scientific Reports*, *3*, 1735.

117) Pentland, A. (2014). *Social physics: How good ideas spread -
The lessons from a new science*. New York, NY: Penguin.

118) Pickard, G., Pan, W., Rahwan, I., Cebrian, M., Crane, R., Madan,
A., & Pentland, A. (2011). Time-critical social mobilization.
*Science*, *334*(6055), 509-512.

119) Mani, A., Rahwan, I., & Pentland, A. (2013). Inducing peer pressure
to promote cooperation. *Scientific reports*, *3*, 1735.

120) Tsao Foundation. (n.d.). Integrated Care System. Retrieved October
21, 2016, from http://tsao-foundation.org/what-we-do/comsa/
integrated-care-system

121) Stanford Vaden Health Center (n.d.). Peer health educators.
Retrieved October 21, 2016, from https://vaden.stanford.edu/about/
training-and-service-opportunities/peer-health-educators

122) Brown, J. S., & Duguid, P. (2000). *The social life of information*. Cambridge, MA: Harvard Business Press.

123) Stanford University (2016, September). "Artificial intelligence and life in 2030." One hundred year study on artificial intelligence: Report of the 2015-2016 study panel. Stanford, CA: Stanford University. Retrieved October 21, 2016, from https://ai100.stanford.edu/2016-report

Stanford University's "Artificial Intelligence and Life in 2030" highlighted that "Contrary to the more fantastic predictions for AI in the popular press, the Study Panel found no cause for concern that AI is an imminent threat to humankind."

124) Lee, C. E., & Satku, K. (Eds.). (2015). *Singapore's health care system: What 50 years have achieved*. Singapore: World Scientific.

125) Ministry of Health, Singapore. (1993, October 22). Affordable health care: A white paper. Retrieved October 21, 2016, from https://www.moh.gov.sg/content/dam/moh_web/Publications/Reports/1993/Affordable_Health_Care.pdf

"The key question of health care financing is who pays for it – individuals, the Government, insurance companies, or employers. This belief is mistake, because no matter who pays for health care in the first instance, ultimate the burden is borne by the people themselves. If the Government pays, it must collect taxes from the people to do so. If insurance companies pay, they must collect premiums from those who are insured. If employers pay, they must add this to their wage costs, and trade it off against other components of the wage package. The real issue is therefore not who pays, but which system best encourages people to use health care economically, and encourages health care providers to minimise costs, inefficiency and over-servicing."

126) Speech by Minister for Health, Mr Gan Kim Yong, at the MOH Committee of Supply Debate 2016. (2016, April 13). Retrieved October 24, 2016, from https://www.moh.gov.sg/content/moh_web/home/pressRoom/speeches_d/2016/speech-by-minister-for-health--mr-gan-kim-yong--at-the-moh-commit.html

127) Crow, D. (2016, July 18). Biotech dives into Sardinia gene pool for secret of long life. Retrieved October 24, 2016, from https://www.ft.com/content/22684864-4b86-11e6-8172-e39ecd3b86fc

128) IBM Research. (n.d.). Computational creativity. Retrieved October 21, 2016, from http://www.research.ibm.com/cognitive-computing/computational-creativity.shtml

129) IBM Research. (2013, November 22). Computational Creativity [Video file]. Retrieved October 21, 2016, from https://www.youtube.com/watch?v=mr-1JAnairs

130) Jain, A., & Bagler, G. (2015). Spices form the basis of food pairing in Indian cuisine. arXiv preprint arXiv:1502.03815.

131) Perez, S. (2015, August 3). Soylent debuts its ready-to-drink meal replacement shake. Retrieved October 21, 2016, from https://techcrunch.com/2015/08/03/soylent-debuts-its-ready-to-drink-meal-replacement-drink-soylent-2-0/

132) Devitt, E. (2016, August 23). As lab-grown meat and milk inch closer to U.S. market, industry wonders who will regulate? Retrieved October 21, 2016, from http://www.sciencemag.org/news/2016/08/lab-grown-meat-inches-closer-us-market-industry-wonders-who-will-regulate

133) Goldberg, H. (2014, July 11). People are still totally confused about local vs. organic. Retrieved October 21, 2016, from http://time.com/2970505/organic-misconception-local/

134) Agri-food & Veterinary Authority of Singapore. (n.d.). Supporting Local Produce. Retrieved October 21, 2016, from http://www.ava.gov.sg/explore-by-sections/food/singapore-food-supply/supporting-local-produce

135) Lim, J. (2015, May 25). Appetite for local produce growing. The Straits Times. Retrieved October 21, 2016, from http://www.straitstimes.com/singapore/appetite-for-local-produce-growing

136) Yotka, S. (2015, August 13). Exclusive! Inside the museum at FIT's newly remodeled archives with Valerie Steele. Retrieved October 21, 2016, from http://www.vogue.com/13293935/museum-at-fit-clothing-archives-fashion-history/

137) Sherman, L. (2013, November 18). For brands big and small, fashion archives can be a powerful asset. Retrieved October 21, 2016, from https://www.businessoffashion.com/articles/intelligence/for-brands-big-and-small-fashion-archives-can-be-a-powerful-asset

138) Tan, C. (2006). SARS in Singapore-key lessons from an epidemic. *Annals of the Academy of Medicine Singapore*, *35*(5), 345.

139) Tan, C. C. (2005). Public health response: A view from Singapore. In Peiris, M., Anderson, L. J., Osterhaus, A. D. M. E., Stohr, K., & Yuen, K. Y. (Eds.), *Severe acute respiratory syndrome* (pp. 139-164). Oxford: Blackwell Publishing.

140) Lee, C. E., & Satku, K. (2015). Challenges in healthcare. In C.E. Lee, & K. Satku (Eds.). Singapore's health care system: *What 50 years have achieved* (pp. 375-386). Singapore: World Scientific.

141) Mate, K. S. & Compton-Phillips, A.L. (2014, December 15) The antidote to fragmented health care - Ideas and advice. Retrieved

October 21, 2016, from
https://hbr.org/2014/12/the-antidote-to-fragmented-health-care

142) Thomas, J. S., Ee, O. S., Seng, C. K., & Peng, L. H. (2015). A brief
history of public health in Singapore. In C.E. Lee, & K. Satku (Eds.).
Singapore's health care system: What 50 years have achieved
(pp. 33-56). Singapore: World Scientific.

From pp. 41-42:
"Further, cities such as Singapore are key nodes in infectious disease
as they are hubs for national, regional, and global spread; bridge
human and animal ecosystems... The 'networked disease' lens posits
that developed-world cities are key facilitators for the global movement
of pathogens."

143) Horby, P. W., Pfeiffer, D., & Oshitani, H. (2016). Prospects for
emerging infections in East and Southeast Asia 10 years after Severe
Acute Respiratory Syndrome. *Emerging Infectious Diseases, 19*(6),
853-860.

144) Washer, P. (2011). Lay perceptions of emerging infectious diseases:
a commentary. *Public Understanding of Science, 20*(4), 506-512.

145) Lee, C. E., & Satku, K. (Eds.). (2015). Singapore's health care system:
*What 50 years have achieved*. Singapore: World Scientific.

From pp. 155-156:
"Although there is no international consensus... 'allied health
professions'... in Singapore.. refer to healthcare professionals who work
alongside the doctors and nurses in the public healthcare system; for
example, radiographers, physiotherapists, occupational therapists,
clinical psychologists, speech therapists, podiatrists, audiologists,
medical social workers, dieticians, amongst others... [they] are integral
members of various multi-disciplinary healthcare teams."

146) Cuddy, A. (2015, December 12). Your iPhone is ruining your posture — and your mood. Retrieved October 22, 2016, from http://www.nytimes.com/2015/12/13/opinion/sunday/ your-iphone-is-ruining-your-posture-and-your-mood.html

"Technology is transforming how we hold ourselves, contorting our bodies into what the New Zealand physiotherapist Steve August calls the iHunch. I've also heard people call it text neck, and in my work I sometimes refer to it as iPosture."

147) Grossman, C., & McGinnis, J. M. (Eds.). (2011). *Digital infrastructure for the learning health system: the foundation for continuous improvement in health and health care: workshop series summary.* Washington, D.C.: National Academies Press.

148) Osman, F. A. (2015). Healthcare providers' attitudes toward using the technology of smart health cards. *Journal of Ubiquitous Systems & Pervasive Networks, 6*(2), 11-17.

149) World Health Organisation. (2013, November 11). Global health workforce shortage to reach 12.9 million in coming decades. Retrieved October 21, 2016, from http://www.who.int/mediacentre/news/ releases/2013/health-workforce-shortage/en/

| | Number of physicians per 1000 population | Number of physicians per 1000 population | Nurses and midwives per 1000 population | Nurses and midwives per 1000 population |
|---|---|---|---|---|
| | 1998 | 2011 | 1998 | 2011 |
| USA | 2.1 | 2.5 | 9.37 | 9.8 |
| Cambodia | 0.1 | 0.2 | 0.9 | 0.9 |
| Norway | 2.7 | 3.7 | 13.4 | 17.3 |
| Singapore | 1.6 | 1.7 | 5.2 | 5.8 |
| Vietnam | 0.5 | 1.2 | 1.1 | 1.2 |

150) Campbell, J., Dussault, G., Buchan, J., Pozo-Martin, F., Guerra Arias, M., Leone, C., ... & Cometto, G. (2013). A universal truth: no health without a workforce. Geneva, CH: World Health Organisation.

The World Health Organisation (WHO) predicts a global shortage of close to 13 million healthcare workers by 2035. The view from experts is that the "the biggest obstacle to improving health is the lack of health workers."

151) Ziebland, S., Coulter, A., Calabrese, J. D., & Locock, L. (Eds.). (2013). Understanding and using health experiences: improving patient care. New York, NY: Oxford University Press.

152) [Personal interview]. (2016).

Pandemics - A doctor we interviewed told us that because of his experience going through SARS, he has become more committed as a healthcare worker. In general, infectious diseases are usually new and unknown, and we might not be able to tell the difference between a normal human variation from an abnormal one, and we will have to be prepared for the unexpected because we do not know what we do not know.

153) BBC Radio 4. (n.d.). The Reith lectures, Dr Atul Gawande - 2014 Reith lectures. Retrieved October 22, 2016, from http://www.bbc.co.uk/programmes/articles/6F2X8TpsxrJpnsq82hggHW/dr-atul-gawan-de-2014-reith-lectures

In an interesting example of increasing specialisations, in the 2014 BBC Reith Lectures, Dr Atul Gawande describes the following:

"We have made tremendous discoveries, but find it's extremely complex to deliver on them. My mother went for a total knee replacement and I counted the number of people who walked in the room in three days and it was 66 different people. And so the complexity of making 66 people work together..."

154) Monllos, K. (2015, October 5). Why brands like Coca-Cola and Bud Light are making packaging personal. Retrieved October 21, 2016, from http://www.adweek.com/news-gallery/advertising-branding/why-brands-coca-cola-and-bud-light-are-making-packaging-personal-167340

155) Mundy, S. (2016, June 13). India ecommerce growth found in translation. Retrieved October 21, 2016, from https://www.ft.com/content/163dc810-2efe-11e6-bf8d-26294ad519fc

156) Diamonds, P. (2016, January 6). These 11 technologies will go big in 2015. Retrieved October 21, 2016, from http://singularityhub.com/2015/01/06/2015s-11-biggest-new-technologies-to-watch/

157) Hart, B., & Risley, T. R. (2003). The early catastrophe: The 30 million word gap by age 3. *American Educator, 27*(1), 4-9.

158) Marulis, L. M., & Neuman, S. B. (2010). The effects of vocabulary intervention on young children's word learning a meta-analysis. *Review of Educational Research, 80*(3), 300-335.

159) Teng, A. (2016, May 17). Bilingual babies 'learn languages faster'. *The Straits Times*. Retrieved October 21, 2016, from http://www.straitstimes.com/singapore/education/bilingual-babies-learn-languages-faster

160) National University of Singapore. (2016, May 23) Bilingual babies know when the rules don't apply. Retrieved October 21, 2016, from http://www.futurity.org/bilingual-babies-mandarin-1168542-2/

161) IKEA Singapore. (n.d.). Soft toys for education - Kids design for a good cause. Retrieved October 21, 2016, from http://www.ikea.com/ms/en_SG/good-cause-campaign/soft-toys-for-education/kids-design-for-good-cause/index.html

162) Coldewey, D. (2016, August 17). Picture this clothing turns your kid's crayon art into a sweet dress. Retrieved October 21, 2016, from https://techcrunch.com/2016/08/17/picture-this-clothing-turns-your-kids-crayon-art-into-a-sweet-dress/

163) Cadoret, R. J., Yates, W. R., Woodworth, G., & Stewart, M. A. (1995). Genetic-environmental interaction in the genesis of aggressivity and conduct disorders. *Archives of General Psychiatry*, 52(11), 916-924.

164) National Human Genome Research Institute. (2015, August 27). Genome-wide association studies fact sheet. (n.d.). Retrieved October 21, 2016, from https://www.genome.gov/20019523/genomewide-association-studies-fact-sheet/

165) Hudson, S. E. (2014, April). Printing teddy bears: a technique for 3D printing of soft interactive objects. *Proceedings of the SIGCHI Conference on Human Factors in Computing Systems*, 459-468. New York, NY: ACM.

166) Ibid.

167) Guo, W., Xu, C., Wang, X., Wang, S., Pan, C., Lin, C., & Wang, Z. L. (2012). Rectangular bunched rutile TiO2 nanorod arrays grown on carbon fiber for dye-sensitized solar cells. *Journal of the American Chemical Society*, 134(9), 4437-4441.

168) Yang, Z., Deng, J., Sun, X., Li, H., & Peng, H. (2014). Stretchable, wearable dye-sensitized solar cells. *Advanced Materials*, 26(17), 2643-2647.

169) Pan, S., Yang, Z., Chen, P., Deng, J., Li, H., & Peng, H. (2014). Wearable solar cells by stacking textile electrodes. *Angewandte Chemie International Edition*, *53*(24), 6110-6114.

170) Lee, Y.H., Kim, J.S., Noh, J., Lee, I., Kim, H.J., Choi, S., Seo, J., Jeon, S., Kim, T.S., Lee, J.Y. & Choi, J.W. (2013). Wearable textile battery rechargeable by solar energy. *Nano letters*, *13*(11), 5753-5761.

171) Zhang, Z., Yang, Z., Wu, Z., Guan, G., Pan, S., Zhang, Y., Li, H., Deng, J., Sun, B. & Peng, H. (2014). Weaving efficient polymer solar cell wires into flexible power textiles. *Advanced Energy Materials*, *4*(11).

172) Chae, Y., Kim, S. J., Kim, J. H., & Kim, E. (2015). Metal-free organic-dye-based flexible dye-sensitized solar textiles with panchromatic effect. *Dyes and Pigments*, *113*, 378-389.

173) Hang, C. C., Low, T. S., & Thampuran, R. (Eds.). *The Singapore research story*. Singapore: World Scientific.

174) Design Singapore Council. (n.d.). The future of Singapore design-Design 2025. Retrieved October 21, 2016, from https://www.designsingapore.org/who_we_are/why_design/Design2025.aspx

175) Ibid.

176) Design Singapore Council. (2016, February 25). Design 2025 masterplan. Retrieved October 21, 2016, from https://www.designsingapore.org/Libraries/Docs/Design2025Masterplan_v2.sflb.ashx

177) Design Singapore Council. (2016, March 29).A design masterplan to take Singapore to 2025. Retrieved October 21, 2016, from https://www.designsingapore.org/who_we_are/news/design_news/16-03-29/A_Design_Masterplan_to_take_Singapore_to_2025.aspx

178) Design Singapore Council. (2016, March 10). Singapore: A thriving innovation-driven economy and a loveable city – by design. Retrieved October 21, 2016, from https://www.designsingapore.org/who_we_are/media_centre/media_releases/16-03-10/MEDIA_RELEASE_Design_2025.aspx

179) Bird, B. (Director). (2004). The Incredibles [Motion picture]. Pixar. Retrieved October 21, 2016, from http://www.pixar.com/features_films/THE-INCREDIBLES

Machine washable quote inspired by Edna Mode scenes from the animated feature film *The Incredibles*.

# Chapter 7

1) Batty, M. (2013). *The new science of cities*. Cambridge, MA: MIT Press.

2) Ibid.

3) Ibid.

4) Levinson, M. (2016). *The box: How the shipping container made the world smaller and the world economy bigger*. Princeton, NJ: Princeton University Press.

5) Bettencourt, L. M., Lobo, J., Helbing, D., Kühnert, C., & West, G. B. (2007). Growth, innovation, scaling, and the pace of life in cities. In *Proceedings of the National Academy of Sciences, 104*(17), 7301-7306.

Urban scaling sets out how different dimensions of cities - economic performance, innovation, crime rates, infrastructure - are governed by the size of the cities, according to laws of nature and mathematics.

6) Albino, V., Berardi, U., & Dangelico, R. M. (2015). Smart cities: Definitions, dimensions, performance, and initiatives. *Journal of Urban Technology*, 22(1), 3-21.

# Annex A

1) Kunda, G. (2009). *Engineering culture: Control and commitment in a high-tech corporation*. Philadelphia, PA: Temple University Press.

2) Lim, W. K., Sia, S. K., & Yeow, A. (2011). Managing risks in a failing IT project: A social constructionist view. *Journal of the Association for Information Systems*, *12*(6), 414-440.

3) Martin, G., Currie, G., Weaver, S., Finn, R., & McDonald, R. (2016). Institutional complexity and individual responses: delineating the boundaries of partial autonomy. *Organization Studies*. doi: 10.1177/0170840616663241

4) Bertels, S., & Lawrence, T. B. (2016). Organizational responses to institutional complexity stemming from emerging logics: The role of individuals. *Strategic Organization*. doi: 10.1177/1476127016641726

5) Hallett, T. (2010). The myth incarnate: Recoupling processes, turmoil, and inhabited institutions in an urban elementary school. *American Sociological Review*, *75*(1), 52-74.

6) Barley, S. R. (1986). Technology as an occasion for structuring: Evidence from observations of CT scanners and the social order of radiology departments. *Administrative Science Quarterly*, *31*, 78-108.

7) Broadbent, E., Stafford, R., & MacDonald, B. (2009). Acceptance of healthcare robots for the older population: review and future directions. *International Journal of Social Robotics*, *1*(4), 319-330.

8) Jensen, T. B., Kjærgaard, A., & Svejvig, P. (2009). Using institutional theory with sensemaking theory: a case study of information system implementation in healthcare. *Journal of Information Technology, 24*(4), 343-353.

9) Soh, C., & Sia, S. K. (2004). An institutional perspective on sources of ERP package–organisation misalignments. *The Journal of Strategic Information Systems, 13*(4), 375-397.

10) Stewart, A. (1998). *The ethnographer's method.* Thousand Oaks, CA: Sage Publications.

11) Lim, W. K., Sia, S. K., & Yeow, A. (2011). Managing risks in a failing IT project: A social constructionist view. *Journal of the Association for Information Systems, 12*(6), 414-440.

12) Nahar, N., Lyytinen, K., Huda, N., & Muravyov, S. V. (2006). Success factors for information technology supported international technology transfer: Finding expert consensus. *Information & Management, 43*(5), 663-677.

13) Lofland, J., & Lofland, L. H. (2006). *Analyzing social settings.* Belmont, CA: Wadsworth Publishing Company.

14) Ibid.

15) Ibid.

16) Rubin, H. J., & Rubin, I. S. (1995). Keeping on target while hanging loose: Design qualitative interviews. In H. J. Rubin & I. S. Rubin (Eds.), *Qualitative Interviewing: The Art of Hearing Data* (pp. 42-64). Thousand Oaks, CA: Sage Publications.

17) Emerson, R. M., Fretz, R. I., & Shaw, L. L. (1995). *Writing ethno-graphic fieldnotes*. Chicago, IL: University of Chicago Press.

18) Lofland, J., Snow, D.A., Anderson, L., & Lofland, L, H. (2006). *Analyzing social settings*. Belmont, CA: Wadsworth Publishing Company.

19) Stewart, A. (1998). *The ethnographer's method*. Thousand Oaks, CA: Sage Publications.

20) Golden-Biddle, K., & Locke, K. (1993). Appealing work: An investigation of how ethnographic texts convince. *Organization Science*, *4*(4), 595-616.

21) Klein, H. K., & Myers, M. D. (1999). A set of principles for conducting and evaluating interpretive field studies in information systems. *MIS Quarterly, 23*(1), 67-93.

22) Eisenhardt, K. M. (1989). Building theories from case study research. *Academy of Management Review*, *14*(4), 532-550.

23) Emerson, R. M., Fretz, R. I., & Shaw, L. L. (1995). *Writing ethno-graphic fieldnotes*. Chicago, IL: University of Chicago Press.

24) Boyatzis, R. E. (1998). *Transforming qualitative information: Thematic analysis and code development*. Thousand Oaks, CA: Sage Publications.

25) Saldaña, J. (2015). *The coding manual for qualitative researchers*. Thousand Oaks, CA: Sage Publications.

26) Ritchie, J., Lewis, J., Nicholls, C. M., & Ormston, R. (Eds.). (2013). *Qualitative research practice: A guide for social science students and researchers*. Thousand Oaks, CA: Sage Publications.

27) Miaskiewicz, T., & Kozar, K. A. (2011). Personas and user-centered design: How can personas benefit product design processes?. *Design Studies*, *32*(5), 417-430.

28) Pruitt, J., & Grudin, J. (2003, June). Personas: practice and theory. In *Proceedings of the 2003 conference on designing for user experiences* (pp. 1-15). New York, NY: ACM. doi: 10.1145/997078.997089

29) Blythe, M. (2014, April). Research through design fiction: narrative in real and imaginary abstracts. In *Proceedings of the SIGCHI Conference on Human Factors in Computing Systems* (pp. 703-712). New York, NY: ACM. doi: 10.1145/2556288.2557098

30) Weick, K. E. (1989). Theory construction as disciplined imagination. *Academy of Management Review*, *14*(4), 516-531.

# Annex B

1) Robotics Virtual Organization. (2013). A roadmap for U.S. robotics: From internet to robotics. (2013). Retrieved November 1, 2016, from https://robotics-vo.us/sites/default/files/ 2013%20Robotics%20Roadmap-rs.pdf

2) Ibid.

3) Mataric, M. J., Okamura, A., & Christensen, H. (2008). *A research roadmap for medical and healthcare robotics*. Paper presented at the NSF/CCC/CRA Roadmapping for Robotics Workshop (pp. 1-30). Arlington/Washington, DC.

4) Robotics Virtual Organization. (2013). A roadmap for U.S. robotics: From internet to robotics. (2013). Retrieved November 1, 2016, from https://robotics-vo.us/sites/default/files/ 2013%20Robotics%20Roadmap-rs.pdf

5) Bischoff, R., Guhl, T., Wendel, A., Khatami, F., Bruyninckx, H., Siciliano, B., ... & Ibarbia, J.A. (2010, June). euRobotics - Shaping the future of European robotics. In *Proceedings of the International Symposium of Robotics/ROBOTIK* (pp. 728–735).

6) Robotics Virtual Organization. (2013). A roadmap for U.S. robotics: From internet to robotics. (2013). Retrieved November 1, 2016, from https://robotics-vo.us/sites/default/files/2013%20Robotics%20Roadmap-rs.pdf

7) Ibid.

8) Ibid.

9) Cooper, P. (2014, December). National product verification programme: A roadmap for UK manufacturing consultation document. Retrieved November 1, 2016, from http://www.npvp.org.uk/wp-content/uploads/2014/07/NPVP-Roadmap-Report-Version-Final.pdf

10) Robotics Virtual Organization. (2013). A roadmap for U.S. robotics: From internet to robotics. (2013). Retrieved November 1, 2016, from https://robotics-vo.us/sites/default/files/2013%20Robotics%20Roadmap-rs.pdf

11) Bischoff, R., Guhl, T., Wendel, A., Khatami, F., Bruyninckx, H., Siciliano, B., ... & Ibarbia, J.A. (2010, June). euRobotics - Shaping the future of European robotics. In *Proceedings of the International Symposium of Robotics/ROBOTIK* (pp. 728–735).

12) Mataric, M. J., Okamura, A., & Christensen, H. (2008). A research roadmap for medical and healthcare robotics. Paper presented at the NSF/CCC/CRA Roadmapping for Robotics Workshop (pp. 1-30). Arlington/Washington, DC.

13) Bischoff, R., Guhl, T., Wendel, A., Khatami, F., Bruyninckx, H., Siciliano, B., ... & Ibarbia, J.A. (2010, June). euRobotics - Shaping the future of European robotics. In *Proceedings of the International Symposium of Robotics/ROBOTIK* (pp. 728–735).

14) Robotics Virtual Organization. (2013). A roadmap for U.S. Robotics: From internet to robotics. Retrieved November 1, 2016, from https://robotics-vo.us/sites/default/files/ 2013%20Robotics%20Roadmap-rs.pdf

15) Ibid.

16) Ibid.

17) Ibid.

18) Ibid.

19) Ibid.

www.ingramcontent.com/pod-product-compliance
Lightning Source LLC
Chambersburg PA
CBHW051953270326
41929CB00015B/2637